Behavioral Cybersecurity

Applications of Personality Psychology and Computer Science

Behavioral Cybersecurity

Applications of Personality Psychology and Computer Science

Wayne Patterson
Cynthia E. Winston-Proctor

CRC Press
Taylor & Francis Group
Boca Raton London New York

CRC Press is an imprint of the
Taylor & Francis Group, an informa business

CRC Press
Taylor & Francis Group
6000 Broken Sound Parkway NW, Suite 300
Boca Raton, FL 33487-2742

Printed on acid-free paper

International Standard Book Number-13: 978-1-138-61778-0 (Hardback)

Library of Congress Cataloging-in-Publication Data

Names: Patterson, Wayne, 1945- author. | Winston-Proctor, Cynthia E., author.
Title: Behavioral cybersecurity : applications of personality psychology and computer science / Wayne Patterson and Cynthia E. Winston-Proctor.
Description: Boca Raton : Taylor & Francis, CRC Press, 2019.
Identifiers: LCCN 2019000325| ISBN 9781138617780 (hardback : alk. paper) | ISBN 9780429461484 (e-book)
Subjects: LCSH: Computer security. | Computer fraud. | Hacking. | Social engineering.
Classification: LCC QA76.9.A25 P3845 2019 | DDC 005.8--dc 3
LC record available at https://lccn.loc.gov/2019000325

Visit the Taylor & Francis Web site at
http://www.taylorandfrancis.com

and the CRC Press Web site at
http://www.crcpress.com

Dedication

To my partner in life for almost half a century: Savanah Williams. A most incredible woman who inspires me everyday, who has chosen her own incredible paths, and who somehow manages to cope with my difficult challenges; and also to my friends Hamid, Orlando, Martin and Arun, who continue to encourage all my work.

Wayne Patterson

I would like to dedicate this book to my loving family with the hope it inspires the Lindsey generation to pursue solving complex problems by integrating psychology and computer science.

Cynthia E. Winston-Proctor

Contents

Preface

Since the introduction and proliferation of the Internet, problems involved with maintaining cybersecurity have grown exponentially and evolved into many forms of exploitation.

Yet, cybersecurity has had far too little study and research. Virtually all of the research that has taken place in cybersecurity over many years has been done by those with computer science, electrical engineering, and mathematics backgrounds.

However, many cybersecurity researchers have come to realize that to gain a full understanding of how to protect a cyberenvironment requires not only the knowledge of those researchers in computer science, engineering, and mathematics, but those who have a deeper understanding of human behavior: researchers with expertise in the various branches of behavioral science, such as psychology, behavioral economics, and other aspects of brain science.

The authors, one a computer scientist and the other a psychologist, have attempted over the past several years to understand the contributions that each approach to cybersecurity problems can benefit from in this integrated approach that we have tended to call "behavioral cybersecurity."

The authors believe that the research and curriculum approaches developed from this integrated approach provide a first book with this approach to cybersecurity. This book incorporates traditional technical computational and analytic approaches to cybersecurity, and also psychological and human factors approaches.

Among the topics addressed in the book are:

- Introductions to cybersecurity and behavioral science
- Profiling approaches and risk management
- Case studies of major cybersecurity events and "Fake News"
- Analyses of password attacks and defenses
- Introduction to game theory and behavioral economics, and their application to cybersecurity
- Research into attacker/defender personalities and motivation traits
- Techniques for measuring cyberattacks/defenses using cryptography and steganography
- Ethical hacking
- Turing tests: classic, gender, age
- Lab assignments: social engineering, passwords in the clear, privacy study, password meters

The history of science seems to evolve in one of two directions. At times, interest in one area of study grows to the extent that it grows into its own discipline. Physics and chemistry could be described in that fashion, evolving from "natural science." There are other occasions, however, when the underlying approach of one discipline is complemented by a different tradition in a totally separate discipline. The study of computer science can be fairly described as an example of that approach. When the first author of this book was a doctoral student at the University of Michigan in the 1970s, there was no department of computer science. It was soon born as a fusion of mathematics and electrical engineering.

Our decision to create this book, as well as several related courses, arose from a similar perspective. Our training is in computer science and psychology, and we have observed, as have many other scholars interested in cybersecurity, that the problems we try to study in cybersecurity require not only most of the approaches in computer science, but more and more an understanding of motivation, personality, and other behavioral approaches in order to understand cyberattacks and create cyberdefenses.

As with any new approaches to solving problems when they require knowledge and practice from distinct research fields, there are few people with knowledge of the widely separate disciplines, so it requires an opportunity for persons interested in either field to gain some knowledge of the other. We have attempted to provide such a bridge in this book that we have entitled *Behavioral Cybersecurity.*

In this book, we have tried to provide an introductory approach in both psychology and cybersecurity, and as we have tried to address some of these key problem areas, we have also introduced topics from other related fields such as criminal justice, game theory, mathematics, and behavioral economics.

We are hopeful that the availability of this book will provide source material for courses in this growing area of behavioral cybersecurity. We feel that such courses can be offered in computer science curricula, psychology curricula, or as interdisciplinary courses. The section called "Introduction" provides a roadmap for courses that might be called (a) behavioral cybersecurity for computer science and psychology, (b) behavioral cybersecurity for computer scientists with some background in behavioral science, or (c) behavioral cybersecurity for behavioral scientists with some background in computing.

INTRODUCTION

We entered the computer era almost 75 years ago. For close to two-thirds of that time, we could largely ignore the threats that we now refer to as cyberattacks. There were many reasons for this. There was considerable research done going back to the 1970s about approaches to penetrate computer environments, but there were several other factors that prevented the widespread development of cyberattacks. Thus, the scholarship into the defense (and attack) of computing environments remained of interest to a relatively small number of researchers.

Beginning in the 1980s, a number of new factors came into play. First among these was the development of the personal computer, which now allowed for many millions of new users with their own individual access to computing power. Following closely on that development was the expansion of network computing, originally through the defense-supported DARPAnet, which then evolved into the openly available Internet. Now, and with the development of tools such as browsers to make the Internet far more useful to the world's community, the environment was set for the rapid expansion of cyberattacks, both in number and in kind, so the challenge for cybersecurity researchers over a very short period of time became a major concern to the computing industry.

The world of computer science was thus faced with the dilemma of having to adapt to changing levels of expertise in a very short period of time. The first author of this book began his own research in 1980, in the infancy of what we now call cybersecurity, even before the widespread development of the personal computer and the Internet.

In the attempt to try to address the need for an accelerated development of researchers who can address the problems of cyberattacks, our two authors have recognized that in addition to the traditional expertise required in studying such problems—that is, expertise in computer science, mathematics, and engineering—we also have a great need to address the human behavior, in the first place, of persons involved in cyberattacks or cybercrime of many forms, but also in the behavioral aspects of all computer users, for example, those who would never avoid precautions in their life such as locking their doors, but use the name of their significant other, sibling, or pet as a password on their computer accounts.

As a result, we have embarked on this project in order to introduce into the field an approach to cybersecurity that relies upon not only the mathematical, computing, and engineering approaches but also depends upon a greater understanding of human behavior. We have chosen to call this subject area "behavioral cybersecurity" and have developed and offered a curriculum over the past several years that now has evolved into this textbook, which we hope will serve as a guidepost for universities, government, industry, and others that wish to develop scholarship in this area.

This book is being proposed (1) for use in developing cybersecurity curricula, (2) as support for further research in behavioral science and cybersecurity, and (3) to support practitioners in cybersecurity.

Behavioral Cybersecurity provides a basis for new approaches to understanding problems in one of our most important areas of research—an approach, agreed upon by most cybersecurity experts, of incorporating not only traditional technical computational and analytic approaches to cybersecurity, but also developing psychological and human-factor approaches to these problems.

The confluence of external events—the power of the Internet, increasing geopolitical fears of "cyberterrorism" dating from 9/11, a greater understanding of security needs and industry, and economic projections of the enormous employment needs in cybersecurity—has caused many universities to develop more substantial curricula in this area, and the United States National Security Agency has created a process for determining Centers of Excellence in this field.

Undergraduate enrollments have been increasing to full capacity. However, we feel there is still a gap in the cybersecurity curriculum that we decided to address.

Background

At the 1980 summer meeting of the American Mathematics Society in Ann Arbor, Michigan, a featured speaker was the distinguished mathematician the late Peter J. Hilton. Dr. Hilton was known widely for his research in algebraic topology, but on that occasion, he spoke publicly for the first time about his work in cryptanalysis during World War II at Hut 8 in Bletchley Park, the home of the now-famous efforts to break German encryption methods such as Enigma.

The first author was present at that session and has often cited Professor Hilton's influence in sparking interest in what we now call cybersecurity. Hilton at the time revealed many of the techniques used at Bletchley Park in breaking the Enigma code. However, one that was most revealing was the discovery by the British team that, contrary to the protocol, German cipher operators would send the same message twice, something akin to, "How's the weather today?" at the opening of an encryption session. (This discovery was represented in the recent Academy-Award–nominated film *The Imitation Game*.) Of course, it is well known in cryptanalysis that having two different encryptions of the same message with different keys is an enormous clue in breaking a code. Thus, it is not an exaggeration to conclude that a behavioral weakness had enormous practical consequences, as the Bletchley Park teams have been credited with saving millions of lives and helping end the war.

Contemporary Behavioral Issues in Cybersecurity

This one example, as important as it is in our history, is repeated countless times in our current cyberspace environments. Most cybersecurity experts will concur that the greatest challenge to effective security is the weakness in human behavior in compromising the technical approach, and not the strength of a technical solution. The first point relates to the lack of motivation of computer users in creating secure passwords, therefore providing a motivation for those who would profit from weak passwords to hack into computer systems and networks.

Cybersecurity researchers generally agree that our field has made spectacular gains in developing technically secure protocols, but all of the careful research in this regard can be overcome by honest users who for some reason choose easy-to-guess passwords such as their significant other's or spouse's name—or on the other hand, hackers who can find such easy-to-guess passwords.

It is believed that in order to counter the clever but malicious behavior of hackers and the sloppy behavior of honest users, cybersecurity professionals (and students) must gain some understanding of motivation, personality, behavior, and other theories that are studied primarily in psychology and other behavioral sciences.

Consequently, by building a behavioral component into a cybersecurity program, it is felt that this curricular need can be addressed. In addition, noting that while only 20% of computer science majors in the United States are women, about 80% of psychology majors are women. It is hoped that this new curriculum, with a behavioral science orientation in the now-popular field of cybersecurity, will induce more women to want to choose this curricular option.

Course Structure

In terms of employment needs in cybersecurity, estimates indicate "more than 209,000 cybersecurity jobs in the U.S. are unfilled, and postings are up 74% over the past five years."

It is believed that the concentration in behavioral cybersecurity will also attract more women students since national statistics show that whereas women are outnumbered by men by approximately 4 to 1 in computer science, almost the reverse is true in psychology.

Our objective with this textbook is to encourage many more opportunities to study and research the area of cybersecurity through this approach to behavioral cybersecurity. With a new approach to the skill set needed for cybersecurity employment, it is hoped that an expanded pool of students will seek to follow this path.

It has also not escaped our notice that the field of cybersecurity has been less attractive to women. Estimates have shown that even though women are underrepresented in computer science (nationally around 25%), in the computer science specialization of cybersecurity, the participation of women drops to about 10%.

However, with the development of a new path through the behavioral sciences into cybersecurity, we observed that approximately 80% of psychology majors, for example, are female. We hope that this entrée to cybersecurity will encourage more behavioral science students to choose this path, and that computer science, mathematics, and engineering students interested in this area will be more inclined to gain a background in psychology and the behavioral sciences.

We feel that this textbook can be applicable to three types of courses: first, classes where it is expected or required that the students have at least some background in both the computer and behavioral sciences; a second path could be for students who have primarily a computing background and little knowledge or expertise in the behavioral sciences; and third, a path for those students whose background is primarily in the behavioral sciences and only minimally in the computing disciplines. What follows describes three separate approaches to the use of this textbook that we will designate as:

- Behavioral cybersecurity for computer science and psychology
- Behavioral cybersecurity for computer scientists with some background in behavioral science
- Behavioral cybersecurity for behavioral scientists with some background in computing

In the following pages, you will see three chapter selections that may be most appropriate for students with the backgrounds described above. The overall chapters are:

Number	Chapter
0	Preface
1	What Is Cybersecurity?
2	Essentials of Behavioral Science
3	Psychology and Cybersecurity
4	Recent Events
5	Profiling
6	Hack Lab 1: Social Engineering Practice: Who Am I?
7	Access Control
8	The First Step: Authorization
9	Hack Lab 2: Assigned Passwords in the Clear
10	Origins of Cryptography
11	Hack Lab 3: Sweeney Method

12	Hacker Personalities: Case Studies
13	Game Theory
14	Ethical Hacking
15	The Psychology of Gender
16	Turing Tests
17	Personality Tests, Methods, and Assessment
18	Modular Arithmetic and Other Computational Methods
19	Modern Cryptography
20	Steganography
21	Using Cryptography and Steganography in Tandem or in Sequence
22	A Metric to Assess Cyberattacks
23	Behavioral Economics
24	Fake News
25	Potpourri
26	Hack Lab 4: Contradictions in Password Meters
27	Conclusion

In the diagrams below, the chapters that are noted with dotted lines may be omitted for the particular group concerned.

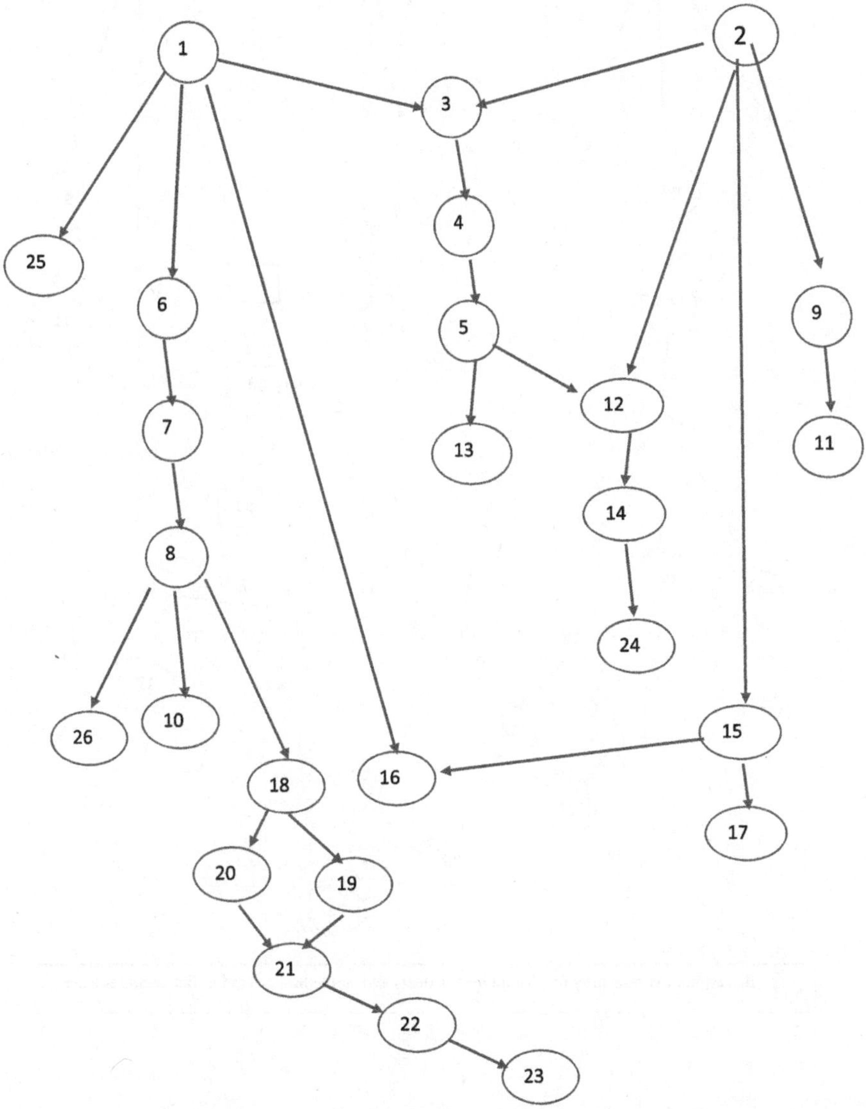

Behavioral Cybersecurity for Computer Science and Psychology

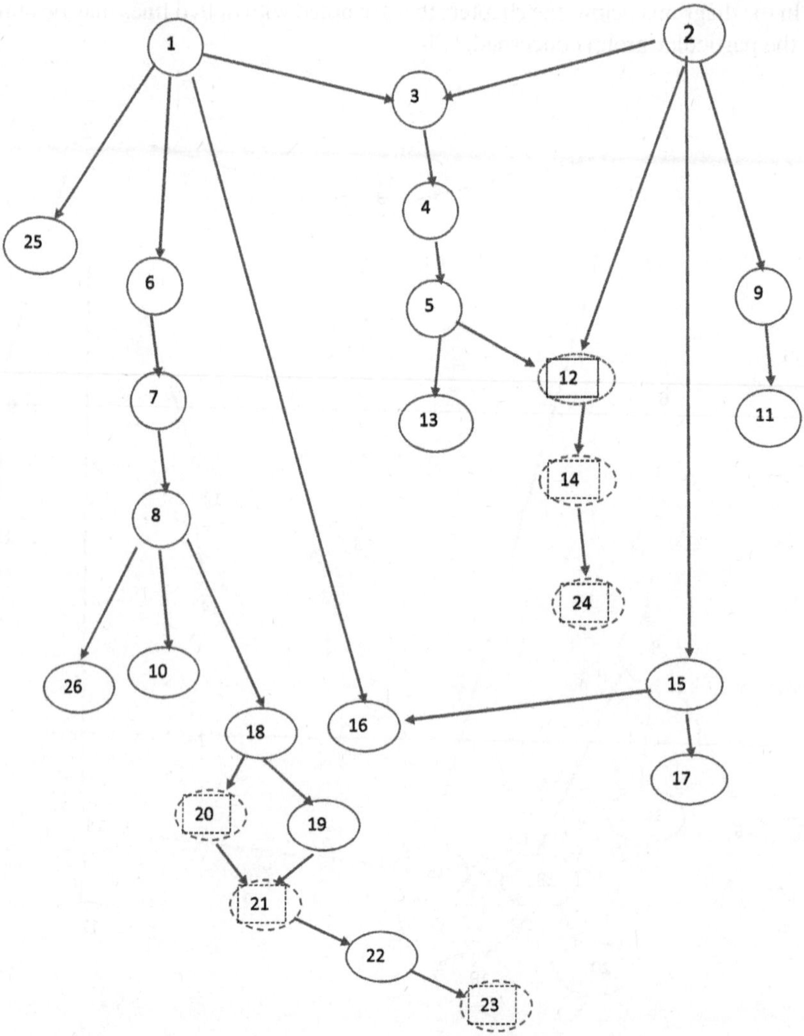

Behavioral Cybersecurity for Computer Scientists with some background in Behavioral Science

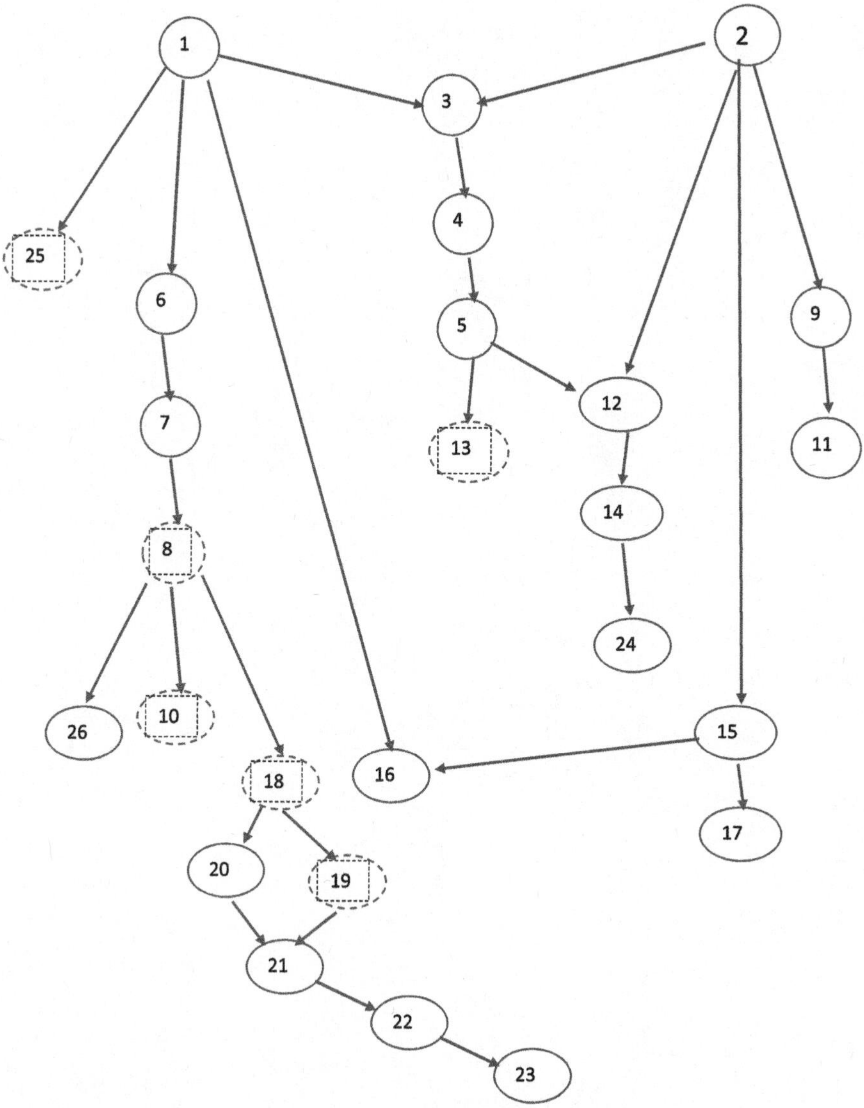

Behavioral Cybersecurity for Behavioral Scientists with some background in Computing

Authors

Dr. Wayne Patterson is a retired professor of computer science from Howard University. He is also currently coprincipal investigator for the National Science Foundation-funded GEAR UP project at Howard, which has supported almost 300 STEM undergrads to do summer research in 15 developing countries. He has also been director of the Cybersecurity Research Center, associate vice provost for Research, and senior fellow for Research and International Affairs in the Graduate School at Howard. He has also been Professeur d'Informatique at the Université de Moncton, chair of the Department of Computer Science at the University of New Orleans, and in 1988, associate vice chancellor for Research there. In 1993, he was appointed vice president for Research and Professional and Community Services, and dean of the Graduate School at the College of Charleston, South Carolina. In 1998, he was selected by the Council of Graduate Schools, the national organization of graduate deans and graduate schools, as the dean in Residence at the national office in Washington, DC. His other service to the graduate community in the United States has included being elected to the presidency of the Conference of Southern Graduate Schools, and also to the Board of Directors of the Council of Graduate Schools. Dr. Patterson has published more than 50 scholarly articles primarily related to cybersecurity; one of the earliest cybersecurity textbooks, *Mathematical Cryptology*; and subsequently three other books. He has been the principal investigator on over 35 external grants valued at over $6,000,000. In August 2006, he was loaned by Howard University to the U.S. National Science Foundation to serve as the Foundation's Program Manager for International Science and Engineering in Developing Countries, and in 2017 was Visiting Scholar at Google.

He received degrees from the University of Toronto (BSc and MSc in Mathematics), University of New Brunswick (MSc in Computer Science), and University of Michigan (PhD in Mathematics). He has also held postdoctoral appointments at Princeton University and the University of California, Berkeley.

Dr. Cynthia E. Winston-Proctor is a widely respected and accomplished narrative personality psychologist and academic. She is professor of Psychology and principal investigator of the Identity and Success Research Laboratory at Howard University. She is also founder of Winston Synergy LLC, a psychology and education consulting firm. Dr. Winston-Proctor earned her BS in psychology from Howard University and her PhD in psychology and education from the University of Michigan. Recognized as an outstanding psychologist, research scientist, and teacher, Dr. Winston-Proctor was awarded the National Science Foundation Early Career Award for scientists and engineers, the Howard University Syllabus of the Year Award, the Howard University Emerging Scholar Award, and a Brown University Howard Hughes Medical Institute Research Professorship. Also, she was elected as a member of the Society of Personology, the oldest and most prominent society for scholars to develop, preserve, and promote theory and research that focuses on the study of individual lives and whole persons. As an academic, she has led the development of curricula across a spectrum of areas including undergraduate education in psychology, behavioral cybersecurity, qualitative inquiry in psychology, healthy living for women, culturally responsive computational thinking, and African ancestry education. Her theory and method development-focused research and education scholarship have resulted in publications in numerous journals and edited books, including *Culture & Psychology*; *Qualitative Psychology*; *Journal of Research on Adolescence*; *Psych Critiques*; *New Directions in Childhood & Development*; the *Oxford Handbook of Cultural Psychology*; and *Culture, Learning, & Technology: Research and Practice.* Dr. Winston-Proctor's professional service includes serving as an editor on the Editorial Board of the American Psychological Association *Journal of Qualitative Psychology*, president of the Society of STEM Women of Color, member of the Board of Directors of the Alfred Harcourt Foundation, and advisor to the Board of Directors of the Howard University Middle School of Mathematics and Science.

1 What Is Cybersecurity?

For the first 40 years or so of the computer era, the question of security was on the one hand widely ignored, and on the other hand relatively simple to address. The reasons, of course, were that far fewer people had access to computing, and also the environment for the computer user was essentially a corporate or university mainframe computer that had no connectivity with other machines that were outside of that corporate environment.

By the mid-1980s, a number of developments began to occur that changed a relatively simple problem to one of massive proportions. In chronological order, events that radically changed the nature of the security problem were:

1. The invention and proliferation of the personal computer in the early 1980s that brought computing power to the individual user.
2. The remarkable PhD thesis by Fred Cohen (1987) that defined the term "computer virus" and demonstrated how such software could completely gain control of the most secure mainframe environment in a matter of a few hours.
3. In 1984, the primary computing society, the Association for Computing Machinery, awarded its Turing Award to two of the developers of the UNIX operating system, Ken Thompson and Dennis Ritchie. Thompson (1984), in his award acceptance speech, posed the challenging problem for programmers of writing a program whose output would be the code of the program itself. Others began to see that such code could be used to create what has been called a computer worm.
4. By the late 1980s, the network ARPAnet, developed much earlier by the U.S. Defense Advanced Research Production Agency (DARPA), started to expand by leaps and bounds, and, with the development of user-friendly software such as browsers, attracted many more people to the use of the Internet, which evolved from ARPAnet.
5. In 1987, the first widespread attack on multiple computers, called the Internet worm, was launched, and it disabled thousands of computers, mostly on university campuses. A Cornell University computer science graduate student, Robert Morris, was charged with and convicted of this crime—he later went on to become a professor of computer science at MIT (Stoll, 1989).
6. On September 11, 2001, the airplane attacks engineered by Osama Bin Laden on the two World Trade Center towers, the Pentagon, and a fourth that crashed near Pittsburgh raised the concerns in the United States and around the world to a new level and foresaw cybersecurity problems.

Since that time, numerous other attacks have led to cybersecurity stories in almost daily headlines. Julian Assange's organization, WikiLeaks, initially won international awards for exposing corruption in organizations and governments. U.S. Army Private Bradley Manning (who later, as a trans woman, changed her name to Chelsea Manning) was able to extract many U.S. government classified documents and make them public via WikiLeaks. Edward Snowden, working with the National Security Agency as a contractor, also obtained classified documents and fled to Russia, where he continues to live.

In addition to these actions by individuals or small organizations, in early 2010, an extremely sophisticated worm called Stuxnet was launched (Nakashima and Warrick, 2012). Spread via Windows, it targeted Siemens software and equipment. It only attacked Siemens Supervisory Control and Data Acquisition System (SCADA) computers. It successfully infected five Iranian organizations related to the Iranian government's processing plants for the enrichment of uranium (either for nuclear power or nuclear weapons, depending on your political perspectives). The result was that the Iranian government indicated that the damage to the enrichment infrastructure cost was the equivalent of $10 million and set the Iranian nuclear program back by an estimated 2 years. The Stuxnet virus was sufficiently sophisticated that most studies of this virus concluded that it could only be built by government levels of organization and investment. It was later discovered that in fact Stuxnet was a joint operation of the United States National Security Agency and Israel's Mossad.

More and more attacks were being discovered, ranging from the trivial (launched by what were often called "spy kiddies") to distributed denial of service (DDoS) attacks designed to bring down websites for perhaps a day at a time—such as happened to MasterCard, Bank of America, the U.S. Department of Justice, and many others. In more recent times, types of attacks called ransomware have been developed, whereby an individual computer may be locked by an external attack until a payment is made in order to free up the attacked computer. Recent examples are the ransomware attacks called WannaCry and Petya, as we will discuss later.

With the explosion of cyberattacks in recent years, the importance of the subject has grown almost without bound. In order to gain an understanding of how to combat these threats, it is necessary to study the subject from a number of points of reference. First of all, it is absolutely necessary to understand the approaches available for the design of a healthy defense strategy. However, it should also be noted that a necessary component of understanding the role of defense is also to understand what possible attack strategies there are. And third, what is often omitted in the study of this field is that the technological approaches described here can be compromised by human behavior, which is why this book seeks to understand both the technological and human behavioral issues that are integral to the study of cybersecurity.

Perhaps the most important historical example of the understanding of the role of human behavior is the breaking of what was thought to have been an unbreakable code, the Enigma code of the German forces in World War II. Although this example essentially predates the invention of the digital computer, the importance is such that it bears repeating. Alan Turing, the brilliant British mathematician and to many the founder of computer science, led the group assigned to break the German Enigma code. The British would obtain daily encrypted messages, but soon learned that the key to the encryption would be changed every day at midnight. Since the number of possible keys was usually in the tens of millions (and their analysis was by hand in the precomputer era), Turing's team was at a loss until it was recognized that certain German cipher operators would begin a day's transmission with the same opening, something akin to "How's the weather today?" (Sony, 2014). It turns out that if a cryptanalyst senses that the same message has been encrypted two different ways, this is a huge step in decrypting an entire message. Once this was realized, the British team was able to break the Enigma messages regularly and continued to do so for the last 4 years of the Second World War without this ability being detected by the Germans. Some historians have concluded that breaking the Enigma code shortened the war by about 2 years and saved on the order of 10 million lives. In essence, the strong cryptographic algorithm of Enigma was defeated by simple human error in following a protocol.

1.1 WHAT IS CYBERSECURITY?

Cybersecurity is a science designed to protect your computer and everything associated with it—the physical environment, the workstations and printers, cabling, disks, memory sticks, and other storage media. But most importantly, cybersecurity is designed to protect all forms of memory, that is, the information stored in your system. Cybersecurity is not only designed to protect against outside intruders, but also both malicious and benign insiders. Of course, the malicious insider often presents the greatest danger, but we also have dangers arising from benign insiders: sharing a password with a friend, failing to back up files, spilling a beverage or food on the keyboard, or natural dangers—the result of a sudden electrical outage, among many other possibilities.

At one time, we could focus on the protection of a single computer. Now, we must consider the entire universe of hundreds of millions of computers to which our machine is connected.

The reason for using the term *cybersecurity* is that at one time, our concern was primarily with a single computer, so if you look back at writings from the 1990s or earlier (Patterson, 1987), you will find that the topics we discussed here tended to be called generically "computer security." But this terminology is clearly out of date, since the number of users whose entire computing environment consists of one machine is dwindling rapidly to zero.

There are three distinct aspects of security: secrecy, accuracy, and availability. Let's consider these in this order.

1.2 SECRECY

A secure computer system does not allow information to be disclosed to anyone who is not authorized to access it. In highly secure systems in government, secrecy ensures that users access only information they're allowed to access. Essentially the same principle applies in industry or academia, since most organizations in society require some level of secrecy or confidentiality in order to function effectively.

One principal difference is that in government systems, including military systems, the rules regarding secrecy may in addition be protected by law.

1.3 ACCURACY: INTEGRITY AND AUTHENTICITY

A secure computer system must maintain the continuing integrity of the information stored in it. Accuracy or integrity means that the system must not corrupt the information or allow any unauthorized malicious or accidental changes to it. Malicious changes, of course, may be affected by an external source, for example, a hacker; however, information may also be changed inadvertently by a less-than-careful user, or also by a physical event such as a fluctuation in an electrical signal.

In network communications, a related variant of accuracy known as authenticity provides a way to verify the origin of data by determining who entered or sent it and by recording when it was sent and received.

1.4 AVAILABILITY

Part of the security requirement for a computer system is availability. In other words, its information must be available to the user at all times.

This means that the computer system's hardware and software keep working efficiently and that the system is able to recover quickly and completely if a disaster occurs.

The opposite of availability is denial of service. Denial of service can be every bit as disruptive as actual information theft, and denial of service has become one of the major threats to the efficient functioning of a computing environment.

1.5 THREATS IN CYBERSECURITY

In describing a scenario for a computing environment that may come under threat, we define three terms:

Vulnerabilities
Threats
Countermeasures

A vulnerability is a point where a system is susceptible to attack. If you were describing a vulnerability in your own home, it might be an unlocked back door.

A threat is a possible danger to the system; for example, a threat could be a person, a thing (a faulty piece of equipment), or an event (a fire or a flood). In the previous example, the threat is a person who exploits the fact that your back door is unlocked in order to gain entry.

Techniques for protecting your system are called countermeasures. To continue the analogy, the countermeasure would consist of locking your back door.

1.6 VULNERABILITIES

In the cybersecurity world, there are many types of vulnerabilities, for example:

Physical vulnerabilities
Natural vulnerabilities

Hardware and software vulnerabilities
Media vulnerabilities
Emanation vulnerabilities
Communications vulnerabilities
Human vulnerabilities

There is a great deal of variation in how easy it is to exploit different types of vulnerabilities. For example, tapping a cordless telephone or a cellular mobile phone requires only a scanner costing perhaps a couple of hundred dollars.

1.7 THREATS

Threats fall into three main categories:

Natural threats
Unintentional threats
Intentional threats

The intentional threats can come from insiders or outsiders. Outsiders can include:

Foreign intelligence agents
Terrorists
Criminals
Corporate raiders
Hackers

1.8 INSIDE OR OUTSIDE?

Although most security mechanisms protect best against outside intruders, many surveys indicate that most attacks are by insiders. Estimates are that as many as 80% of system penetrations are by fully authorized users.

1.9 THE INSIDER

There are a number of different types of insiders: the disgruntled employee, the coerced employee, the careless employee, and the greedy employee. One of the most dangerous types of insiders may simply be lazy or untrained. He or she doesn't bother changing passwords, doesn't learn how to encrypt files, doesn't get around to erasing old disks, doesn't notice a memory stick inserted into the back of the computer, and leaves sensitive printouts in piles on the floor.

1.10 COUNTERMEASURES

There are many different types of countermeasures or methods of protecting information. The fact that in earlier times, our working environment might consist of a single computer—an environment that virtually no longer exists—is the reason that we have retired the term *computer security* and replaced it with *cybersecurity*, which now consists of at least the following needs for countermeasures. Let's survey these methods:

Computer security
Communications security
Physical security

1.11 COMPUTER SECURITY: THEN AND NOW

In the early days of computing, computer systems were large, rare, and very expensive. Those organizations lucky enough to have a computer tried their best to protect it. Computer security was just one aspect of general plant security. Security concerns focused on physical break-ins; theft of computer equipment; and theft or destruction of disk packs, tape reels, and other media. Insiders were also kept at bay. Few people knew how to use computers; thus, the users could be carefully screened. Later on, by the 1970s, technology was transformed, and with it the ways in which users related to computers and data. Multiprogramming, time-sharing, and networking changed the rules.

Telecommunications—the ability to access computers from remote locations—radically changed computer usage. Businesses began to store information online. Networks linked minicomputers together and with mainframes containing large online databases. Banking and the transfer of assets became an electronic business.

1.12 NEW ABUSES

The increased availability of online systems and information led to abuses. Instead of worrying only about intrusions by outsiders into computer facilities and equipment, organizations now had to worry about computers that were vulnerable to sneak attacks over telephone lines and information that could be stolen or changed by intruders who didn't leave a trace. Individuals and government agencies expressed concerns about the invasion of privacy posed by the availability of individual financial, legal, and medical records on shared online databases.

1.13 THE PERSONAL COMPUTER WORLD

The 1980s saw a new dawn in computing. With the introduction of the personal computer (PC), individuals of all ages and occupations became computer users. This technology introduced new risks. Precious and irreplaceable corporate data were now stored on diskettes, which could now be lost or stolen.

As PCs proliferated, so too did PC networks, electronic mail, chat rooms, and bulletin boards, vastly raising the security stakes. The 1980s also saw systems under attack.

1.14 THE FUTURE

The challenge of the next decade will be to consolidate what we've learned—to build computer security into our products and our daily routines, to protect data without unnecessarily impeding our access to it, and to make sure that both products and standards grow to meet the ever-increasing scope of challenge of technology.

REFERENCES

Cohen, F. 1987. Computer viruses: Theory and experiments, *Computers and Security*, 6(1), 22–32.

Nakashima, E. and Warrick, J. *Stuxnet was work of US and Israeli experts, officials say.* Washington Post, June 1, 2012. https://www.washingtonpost.com/world/national-security/stuxnet-was-work-of-us-and-israeli-experts-officials-say/2012/06/01/gJQAlnEy6U_story.html?utm_term=.a2c6db0e1f1a

Patterson, W. 1987. *Mathematical Cryptology*. Totowa, New Jersey: Rowman and Littlefield, 318 pp.

Sony Pictures Releasing. *The Imitation Game* (film), 2014.

Stoll, C. 1989 *The Cuckoo's Egg*. Doubleday, 336 pp.

Thompson, K. 1984. Reflections on trusting trust, *Communications of the ACM*, 27(8), 761–763.

PROBLEMS

All of the sources necessary to answer these questions are available on the Internet if they are not contained in the chapter itself. In particular, the entire book by Stoll can be found online.

1.1 Describe in a paragraph what Fred Cohen proved in his PhD thesis.

1.2 Can you identify the date in the model of the earliest personal computer?

1.3 Read the Kenneth Thompson Turing Award speech. What is his "chicken-and-egg" problem?

1.4 In *The Cuckoo's Egg* by Clifford Stoll, answer the following:

  a. What was the size of the billing error that led to the beginning of the investigation?
  b. What user was responsible for that billing error and why did it get by?
  c. Why was Joe Sventek not a suspect?
  d. What was the clue that the intruder was coming in from outside the labs and not inside?
  e. Who was eventually caught and prosecuted?

1.5 Where are Julian Assange and Edward Snowden currently?

1.6 What is the world population as of December 1, 2018? What is your estimate of the percentage of the world's population with computer availability? Indicate your source.

1.7 Construct a vulnerability/threat/countermeasure example for (a) a theater; (b) a farm.

1.8 What are the 10 countries with the greatest Internet usage by percentage of the population who are Internet users?

2 Essentials of Behavioral Science

Understanding human behavior is integral to the study of cybersecurity. Without a human actor, virtually all cybersecurity issues would be nonexistent. Within computer science and engineering, human-factor psychology is the most common psychological subfield used to solve problems. Human-factor psychology is a scientific discipline that studies how people interact with machines and technology to guide the design of products, systems, and devices that are used every day, most often focusing on performance and safety (Bannon, 1991; Bannon and Bodker, 1991; Salvendy, 2012). Sometimes human-factor psychology is referred to as ergonomics or human engineering. We extend this human-factor psychology focus to include behavioral science generally and personality psychology more specifically to develop theoretical and methodological formulations to solve cybersecurity problems. Personality psychology offers a theoretical and methodological approach that is complementary to human-factor psychology as an integral element of cybersecurity. The purpose of this chapter is to describe dimensions of human personality to advance the argument that a personality analytic approach to behavioral cybersecurity that centers on understanding the attacker's personality can inform defender decision-making in developing a defense strategy. Personality psychology provides a robust theoretical framework for describing differences in individual behavior, while also offering an integrative conceptual framework for understanding the whole person.

2.1 WHAT IS BEHAVIORAL SCIENCE?

Within the context of the history of psychology, behavioral science is a relatively new way to characterize the field of psychology that does so by characterizing itself as the organized study of human and animal behavior through controlled systematic structure (Adhikari, 2016). The American Psychological Association (APA) characterizes the field of psychology as a "diverse discipline," but with nearly boundless applications to everyday life. In part, the field's diversity is because it includes several subdisciplines (e.g., developmental psychology, social psychology, organizational psychology, personality psychology, educational psychology, clinical psychology) and over time it has also situated itself as both a natural science and social science. This positioning of the field at the nexus of both natural and social science is in part due to the history of the field of psychology.

The history of the field of psychology tells the story of a field that has traversed multiple disciplines during its early formation. Until the 1870s, psychology was a domain within the field of philosophy. In 1872 within the United States, William James was appointed instructor in physiology at Harvard College and only later in 1875 taught the first American course in psychology. In 1879 within Germany, psychology developed into an independent scientific discipline through the experimental work of Wilhem Wundt. As a natural science, the field of psychology has pursued questions that have tended to be more focused on understanding the physical world in terms of observable physical behavior (vs. covert behavior of cognition). Historically, methods of observation and empiricism have been used by psychologists to describe psychology as a natural science. As a social science, psychology has evolved to explicitly incorporate the idea that human behavior emerges in the social context of a human society characterized by social dynamics of behavior, culture, and institutions.

In this chapter, we describe dimensions of human personality to advance the argument that a personality analytic approach to behavioral cybersecurity that centers on understanding the attacker's personality can inform defender decision-making in developing a defense strategy. Personality psychology provides a valuable framework for describing differences in individual behavior, while also offering a conceptual framework for understanding the whole person. Each of the three dimensions of human personality (i.e., personality traits, personality characteristic adaptations, and narrative identity) has varying levels of connection to culture.

2.2 WHY PERSONALITY PSYCHOLOGY?

2.2.1 Theoretical Orientation: What Do We Know When We Know a Person?

In a highly influential article, Carlson (1971) raised the question, "Where is the person in personality research?" In so doing, he criticized the field for seeming to retreat from its original mission envisioned by the field's founders (Allport, 1937; Murray, 1938) to provide *an integrative framework for understanding the whole person*. Carlson argued that this is a big problem since the field is uniquely positioned compared to other subfields of psychology to focus on human individuality. This is in contrast to developmental psychology—the goal is to understand the stages and milestones of development; social psychology—the goal is to understand social interactions, dynamics, and cognitions (attitudes); and clinical psychology—the goal is to understand psychological disorders that people experience in their daily living. Personality psychology's mission to focus on the whole person is aligned with what later came to be known as personology. Henry Murray (1938) characterized the study of personology across the social sciences as theory and research that focus on the study of individual lives and whole persons.

Several decades later, McAdams (1995) raised a similar question but with more direct import to the theoretical and empirical canon of knowledge that has developed since the time of Carlson's (1971) article. The title of McAdams's (1995) article was "What Do We Know When We Know a Person?" Over time, from this initial question, McAdams and his colleagues developed a series of scholarly articles and books that have had a profound impact on how personality psychologists have come to theorize, study, and apply personality psychology.

In his book *The Art and Science of Personality Development*, Dan McAdams (2015), a leading scholar of conceptual development in the field of personality psychology, translated the body of theory and research in personality psychology to answer the central question of personality psychology (i.e., what do we know when we know a person?). He asserted that across developmental time, we know an actor that first has a presentation style (temperament) that gradually morphs into the basic dispositional traits of personality; an agent with a dynamic arrangement of evolving motives, goals, values, and projects; an author seeking to figure out who they are and how they are to live, love, and work in adult society in which they are embedded in terms of the social context and culture. Prior to that, in a seminal article published in *American Psychologist* entitled "A New Big Five: Fundamental Principles for an Integrative Science of Personality," McAdams and Pals (2006) developed an integrative framework for personality in which they conceived of personality as "(a) an individual's unique variation on the general evolutionary design for human nature, expressed as a developing pattern of (b) dispositional traits, (c) characteristic adaptations (e.g., motives, goals, social intelligence), and (d) self-defining life narratives, complexly and differentially situated (e) in culture and social context" (p. 204).

2.2.2 Personality Traits and the "Social Actor": The Having Side of Personality

Personality traits are a person's broad, decontextualized, and relatively enduring basic behavioral tendencies or dispositions. Some personality scholars have characterized personality traits as an element of personality that another person could interpret if he/she did not know a person very well. In this same vein of thinking, McAdams (1995) characterized personality traits as sketching an outline of the person; it is in a sense, accordingly to him, the psychology of a stranger whereby you can size up a person upon the first meeting.

To elaborate on the conceptualization of personality traits, it is important to highlight the meaning of each of the descriptors. In terms of personality traits being broad and decontextualized, personality psychologists mean that personality traits tend to be expressed across most situations. During several decades in the field, Mischel (1968, 1973) sparked what came to be known in as the person-situation debate. This debate eroded, for that time, some of the prominence of the personality trait as an important descriptor and predictor of human behavior. In essence, what was debated among psychologists was whether human behavior is more contingent on situation or more cross-situationally consistent (i.e., traitlike). This debate spurred a vast amount of scientific research on personality traits, including research on personality traits predicting psychological outcomes (e.g., Barrick and Mount, 1991; Diener et al., 1992a,b), stability and heritability of personality traits (e.g., Bouchard et al., 1990), and association between personality traits and brain function (e.g., LeDoux, 1996). As a result of this vast corpus of research, personality traits have since dominated the study of human personality in the field of psychology.

To describe personality traits as relatively enduring means that these traits have what personality psychologists call longitudinal consistency (McCrae and Costa, 1999). In other words, a person's personality traits are relatively stable and unchanging across the entire life span (Conley, 1985; McCrae and Costa, 1997, 1999; Roberts and DelVecchio, 2000). Also, it is the dimension of personality that scientists have discovered to be about 50% due to heredity (Bouchard et al., 1990).

The Big Five Personality Trait theory (McCrae and Costa, 1997, 1999; Goldberg, 1993; John and Srivastava, 1999) is generally considered by psychologists the most scientifically valid and robust model of five classes of personality traits. These personality traits are extroversion, neuroticism, openness, conscientiousness, and agreeableness. Each personality trait includes distinguishable components that personality psychologists use as trait adjectives and often refer to in personality trait research as facets of each of the five classes of personality traits.

Personality psychologists have described these five major traits in the following way (John and Srivastava, 1999; McCrae and John, 1992):

Extroversion: The personality trait extroversion includes the following characteristics: (1) activity level (e.g., energetic), (2) dominance, (3) sociability, (4) expressiveness (e.g., showing off), and (5) positive emotionality (spunky). A person high on the personality trait extroversion would be described as a person who is active, assertive, sociable, adventurous, and enthusiastic. A person low on extroversion would be described as more introverted, reserved, and submissive to authority.

Neuroticism: The personality trait neuroticism includes the following characteristics: (1) anxious, (2) self-pitying, (3) hostile. A person high on neuroticism would be described as having the behavioral tendency to experience strong emotional states such as sadness, worry, anxiety, and fear, as well as overreacting to frustration and being overly irritable, self-conscious, vulnerable, and impulsive. A person low on neuroticism would be described as calm, confident, contented, and relaxed.

Conscientiousness: The personality trait conscientiousness includes the following characteristics: (1) competent (e.g., efficient), (2) orderly, (3) dutiful, (4) self-disciplined, and (5) deliberate. A person high on conscientiousness would be described as a person who is dependable, organized, planful, responsible, and thorough. A person low on conscientiousness would be described as irresponsible and easily distracted.

Agreeableness: The personality trait agreeableness includes the following characteristics: (1) warmth, (2) tendermindedness (e.g., kindheartedness), (3) altruism (e.g., generosity), (4) modesty, and (5) trust. A person high on agreeableness would be described as a person who is appreciative, forgiving, kind, and sympathetic. A person low on agreeableness would be described as shy, suspicious, and egocentric.

Openness: The personality trait openness includes the following characteristics: (1) aesthetically reactive, (2) introspective, (3) values intellectual matters. A person high on openness would be described as a person who is artistic, curious, imaginative, insightful, and original and has wide interests. A person low on openness would be described as a person who has a pronounced sense of right and wrong.

Research on personality traits has discovered their association with a number of psychological processes and outcomes. The amount of research on each trait varies widely, with extroversion overwhelmingly receiving the most attention and openness the least. Extroversion is associated with more social interaction on a daily basis (Srivastava et al., 2008), creating a positive social environment for others (Eaton and Funder, 2003); higher levels of social competence (Argyle and Lu, 1990); more social goals (King, 1995); greater striving for intimacy (Roberts and Robins, 2004); strong interest in making money in business (Roberts and Robins, 2000); interest in sales, marketing, and people-oriented professions (Diener et al., 1992a); stronger conditioning for rewards and incentives (Matthews and Gilliland, 1999); better performance at timed tasks, but poorer accuracy (Wilt and Revelle, 2009); and superior performance in divided attention tasks and multitasking (Lieberman and Rosenthal, 2001). Openness is associated with seeking out new and varied experiences (McCrae and Costa, 1997). The personality trait of neuroticism has been discovered to be associated with individuals perceiving threats in the social environment; bad interpersonal experience; and negative outcomes in life including higher divorce rates, poorer health, and increased risk or heart disease and other life-threatening illnesses (Ozer and Benet-Martínez, 2006; Smith, 2006). Both the trait of conscientiousness and that of agreeableness have been associated with secure attachment relationships (Noftle and Shaver, 2006); investment in family roles (Lodi-Smith and Roberts, 2007, 2012); low levels of conflict in personal and family relationships (Asendorpf and Wilpers, 1998); investment in work roles (Lodi-Smith and Roberts, 2007, 2012); academic achievement in college (Corker et al., 2012); career stability (Laursen et al., 2002); difficulty coping with unemployment (Boyce et al., 2010); lower earnings, especially among men (Judge et al., 2012); and longevity (Friedman, 2000, 2008).

2.2.3 Personality Characteristic Adaptations and the Motivated Agent: The Psychology of the Doing Side of Personality

Adaptive functioning is a key feature of human capacity. It is also essential in how personality operates within the second dimension of human personality: characteristic adaptations. McCrae and Costa (1997) were the first theorists to coin the term "characteristic adaptations" to refer to a cluster of personality processes that are highly contextualized. McAdams and Pals (2006) elaborated on their conceptualization of this second dimension of personality by specifying that personality characteristic adaptations are contextualized facets of psychological individuality that speak to motivational, cognitive, and developmental concerns in personality that can adapt to social roles, situations, and developmental periods (e.g., childhood, adolescence, adulthood). Psychologists have invested enormous attention in trying to understand personality characteristic adaptations in the form of human motivation and goals.

2.2.3.1 Human Motivation and Goals: Needs and Strivings

Across time, a key question that has directed motivation and goal theory and research has been "What energizes and directs human behavior?" Psychologists have answered this question for more than 100 years by developing theories that answer it in myriad ways, including the following most common answers: behavior is energized and directed by the same drives that shape behavior of animals; unconscious drives and needs (Freud, 1961); the pursuit of pleasure and avoidance of pain (Freud, 1961);

a hierarchy of needs that begins with meeting physiological needs and ends with self-actualization (Maslow, 1968); problems that human beings have evolved to solve like finding food, locating shelter, attracting and selecting a mate, procreating, etc. (Buss, 1995); self-determination (Deci and Ryan, 1985, 1991) stemming from needs of autonomy, competence, and relatedness afforded by opportunities and constraints within the social environment; and a diversity of needs that include achievement, intimacy, and power (McClelland, 1961, 1985).

As an adaptive characteristic of human personality, theoretically a person will show consistent motivation trends over many decades of life, while also demonstrating significant developmental change within various social roles and situations. Among the most widely studied aspects of human motivation and goals are power motivation, achievement motivation, social motivation, and personalized goals.

2.2.3.2 Power Motivation: Striving for Power

Power motivation is the desire to have an impact on other people, to affect their behavior or emotions, and includes overlapping concepts such as influence, inspiration, nurturance, authority, leadership, control, dominance, coercion, and aggression (Winter, 1992). Winter (1992) suggested that "although these concepts have varying connotations of legitimacy, morality, and implied reaction of the other person (the 'target' of power), they share a core meaning of one person's capacity to affect the behavior or feelings of another person. The idea that humans have a drive for power or 'will to power' is prominent in many cultural and intellectual traditions" (p. 302).

2.2.3.3 Social Motivation: The Striving for Affiliation

Social motivation is the desire to interact with others and gain a sense of affiliation or relatedness. Psychologists have theorized about various forms of social motivation, including belongingness, intimacy motivation, and implicit theories.

Baumeister and Leary (1995) conceptualized belongingness as the human need to form and maintain strong, stable interpersonal relationships. Belongingness is facilitated by the psychological mechanisms of enhancement (and protection) of the self, trusting, understanding, and controlling (Fiske, 2010). These psychological mechanisms facilitate a person being able to effectively function in social groups. Among the core elements of belongingness is the need for affiliation. Heckhausen (1991) described these aspects of social bonding in terms of various forms of social affiliation that include love of family (e.g., children, spouse, parents) and friendship, as well as seeking and maintaining strong relationships with unfamiliar others of the same gender and similar age. Intimacy motivation focuses on the motive to pursue warm, close, and communicative interactions with those with whom an individual has a one-on-one relationship (McAdams, 1985). Another form of social motivation is implicit theories that orient a person to specific drives (Dweck and Leggett, 1988).

2.2.3.4 Achievement Motivation: The Striving for Achievement

Achievement motivation is the reoccurring desire to perform well and strive for success (McClelland et al., 1953). Aspects of achievement motivation include overlapping concepts such as self-efficacy, self-concept of ability, achievement values, self-determination, intrinsic motivation, and extrinsic motivation (see Mayer et al., 2007).

2.2.3.5 Personalized Goals

Personalized goals are a form of what psychologists refer to as personalization of human motivation. Two dimensions of personalized goals are personal strivings and personal projects. Personal strivings are characteristic types of goals that individuals try to achieve through their everyday behavior (Emmons, 1986; Little, 1999). Personal projects can range from goals that are "one-off" pursuits, routine acts, or personal strivings (e.g., exercise), to overarching commitments of a lifetime (e.g., solve problem of world hunger) (Little, 1983).

Research on human motivation and goals has discovered its association with a number of psychological processes and outcomes. Among women and men high in responsibility, Winter and Barenbaum (1985) found that power motivation predicts responsible social power actions, but among women and men low in responsibility, power motivation predicts a variety of profligate, impulsive actions (e.g., aggression, drinking). Achievement motivation had been found to be associated with performance, memory, learning, and perception. Psychologists have revealed that social motivation is related to positive psychological adjustment and well-being (Baumeister and Leary, 1995). Personalized goals contribute to a person's psychological well-being. More specifically, people tend to be happiest when they are making progress toward pursuit of their personal goals, especially when these goals are consistent with more general motivational trends in their personality (i.e., achievement, power, and social motivation).

2.2.4 Narrative Identity and "The Autobiographical Author": Creating Narratives to Live By

People create in their minds a personal life story. Personality psychologists refer to this as a person's narrative identity or an internalized and evolving narrative of self (Winston, 2011). The psychological process of autobiographical narration in which a person engages to cultivate narrative identity is largely unconscious and selective. In so doing, personality psychologists have explained that the person engages in reinterpretation of the past, perceived present, and anticipated future (McAdams, 2001; Singer, 2004, 2005).

The psychological function of narrative identity and life story construction is to provide a person with a sense of purpose, self-coherence, and identity continuity over time. Pressing identity questions that psychologists have identified that people pursue are, "Who am I?" (Erikson, 1968) and "How do I find unity and purpose in life?" (McAdams, 2001) These are the key questions about narrative identity that personality psychologists pursue. In so doing, they focus on theory and research to describe, explain, and uncover a person's internalized and evolving narrative of self. In his seminal identity-statuses theory, Erikson (1963) explained that young adults struggle to develop an identity through traversing a series of psychosocial identity stages focused on love, work, and commitment, with the most developed stage being achieved identity. McAdams (1985) elaborated on Erikson's identity theory with his introduction of the life-story theory of identity in which he reconceptualized the psychosocial journey as being a pursuit to articulate and internalize a story about the self that integrates multiple roles in the person's life, as well as selective life experiences across the life course (Hammack, 2008; McAdams and Manczak, 2011; McAdams, 2001; McLean et al., 2007; Winston-Proctor, 2018).

Based on theoretical and methodological advances in the study of narrative identity, psychologists have identified several cognitive processes that are implicated in a person's discovery of his/her internalized and evolving narrative of self across the life course. The two most important of these processes are narrative processing and autobiographical reasoning. Narrative processing is primarily/principally about the narrative structure of a person's autobiographical memories of experience, while autobiographical reasoning is the meaning-making narrative mental operations (Singer and Bluck, 2001).

In an analysis of the narrative identity findings within personality psychology, as well as developmental psychology and other related fields, McAdams and McLean (2013) identified the following narrative identity themes: agency, communion, redemption, contamination, exploratory narrative processing, and coherent positive resolution.

Personality psychologists have conceptualized autobiographical memories as serving as building blocks for narrative identity (Singer et al., 2013). Autobiographical memories are part of a larger self-memory system. A prominent model of this larger self-memory system is Conway and Pleydell-Pearce's (2000) self-memory system model. This model describes how this memory system contains an autobiographical knowledge base and current goals of the working self. The goal structure of the working self is vital in both encoding and retrieval of autobiographical memories. Some scholars have suggested that autobiographical memories are primarily records of failure and success in goal attainment (e.g., Conway, 1992). Within this self-memory system, autobiographical memories contain knowledge at different levels of specificity: life time period, specific event, general event (Conway and Pleydell-Pearce, 2000).

In sum, the key concepts that are essential for understanding the narrative identity dimension of personality for describing differences in individual behavior and understanding the whole person are as follows:

Narrative Identity: An internalized and evolving narrative of self that is a developmental layer of human personality (McAdams and Pals, 2006; McLean, 2008; Singer, 2004).

Narrative Processing: The construction of storied accounts of life experiences that are dependent on vivid imagery, familiar plot structures, and archetypal characters living within the context of predominant cultural themes or conflicts (Singer and Bluck, 2001).

The Self-Memory System: An autobiographical memory system that contains an autobiographical knowledge base, which is sensitive to cues and patterns of activation based on emotions and goals of the working self (Conway and Pleydell-Pearce, 2000).

Self-Defining Memories: A type of autobiographical memory that is vivid, affectively intense, and well rehearsed. They connect to a person's other significant memories that share their themes and narrative sequences. Singer

> et al. (2013) explain that autobiographical memories also reflect individuals' most enduring concerns (e.g., achievement, intimacy, spirituality) and/or unresolved conflicts (e.g., sibling rivalry, ambivalence about a parental figure, addictive tendencies).

Research on narrative identity has discovered its association with a number of psychological processes and outcomes. In a review of recent narrative identity studies, McAdams and McLean (2013) found that researchers have primarily focused on the relationship between narrative identity and both psychological adaptation and development. From this body of research, they concluded that individuals with a narrative identity that finds "redemptive meaning in suffering and adversity and who construct life stories that feature themes of personal agency and exploration, tend to enjoy higher levels of mental health, well-being, and maturity" (McAdams and McLean, 2013; p. 233). Relatedly, Singer et al. (2013) articulated how health narrative identity combines narrative specificity with adaptive meaning-making to achieve insight and well-being.

2.3 CONCLUSION

A defender of a cyberattack could benefit from having knowledge of the theoretical conceptualizations that behavioral scientists use to determine a person's personality traits, personality characteristic adaptations (e.g., motives, goals), and narrative identity as tools at his or her disposal in order to defeat the attack in question. Gaining knowledge about the theoretical conceptualization of personality used by personality psychologists allows those interested in behavioral cybersecurity to explore on their own trends in personality development that derive from ongoing scientific research. Also, with this knowledge, behavioral cybersecurity scholars and practitioners can explore how these dimensions of personality can be revealed from clues manifested from observations drawn from particular cybersecurity cases and scenarios.

REFERENCES

Adhikari, D. 2016. Exploring the differences between social and behavioral science. *Behavioral Development Bulletin*, 21(2), 128–135.

Allport, G. W. 1937. *Personality: A Psychological Interpretation*. New York: Holt, Rinehart and Winston.

Argyle, M. and Lu, L. 1990. The happiness of extraverts. *Personality and Individual Differences*, 11(10), 1011–1017.

Asendorpf, J. B. and Wilpers, S. 1998. Personality effects on social relationships. *Journal of Personality and Social Psychology*, 74(6), 1531–1544.

Bannon, L. J. 1991. From human factors to human actors: The role of psychology and human-computer interaction studies in system design. In J. Greenbaum and M. Kyng (Eds.), *Design at Work: Cooperative Design of Computer Systems* (pp. 25–44). Hillsdale, NJ: L. Erlbaum Associates.

Bannon, L. J. and Bodker, S. 1991. Beyond the interface: Encountering artifacts in use. In J. M. Carroll (Ed.), *Designing Interaction: Psychology at the Human Computer Interface* (pp. 227–253). New York: Cambridge University Press.

Barrick, M. R. and Mount, M. K. 1991. The Big Five personality dimensions and job performance: A meta-analysis. *Personnel Psychology*, 44, 1–26.

Baumeister, R. F. and Leary, M. R. 1995. The need to belong: Desire for interpersonal attachments as a fundamental human motivation. *Psychological Bulletin*, 117(3), 497–529.

Bouchard, T. J., Jr., Lykken, D. T., McGue, M., Segal, N. L., and Tellegen, A. 1990. Sources of human psychological differences: The Minnesota Study of Twins Reared Apart. *Science*, 250, 223–228.

Boyce, C. J., Wood, A. M., and Brown, G. D. A. 2010. The dark side of conscientiousness: Conscientious people experience greater drops in life satisfaction following unemployment. *Journal of Research in Personality*, 44(4), 535–539.

Buss, D. M. 1995. Psychological sex differences: Origins through sexual selection. *American Psychologist*, 50(3), 164–168.

Carlson, R. 1971. Where is the person in personality research? *Psychological Bulletin*, 75(3), 203–219.

Conley, J. J. 1985. Longitudinal stability of personality traits: A multitrait–multimethod–multioccasion analysis. *Journal of Personality and Social Psychology*, 49, 1266–1282.

Conway, M. A. 1992. A structural model of autobiographical memory. In M. A. Conway, D. C. Rubin, H. Spinnler and E. W. A. Wagenaar (Eds.), *Theoretical Perspectives on Autobiographical Memory* (pp. 167–194). Dordrecht, the Netherlands: Kluwer Academic.

Conway, M. A. and Pleydell-Pearce, C. W. 2000. The construction of autobiographical memories in the self-memory system. *Psychological Review*, 107(2), 261–288.

Corker, K. S., Oswald, F. L., and Donnellan, M. B. 2012. Conscientiousness in the classroom: A process explanation. *Journal of Personality*, 80(4), 995–1028.

Deci, E. L. and Ryan, R. M. 1985. *Intrinsic Motivation and Self-Determination in Human Behavior*. New York: Plenum.

Deci, E. L. and Ryan, R. M. 1991. A motivational approach to self: Integration in personality. In R. Dienstbier (Ed.), *Nebraska Symposium on Motivation: Vol. 38. Perspectives on Motivation* (pp. 237–288). Lincoln, NE: University of Nebraska Press.

Diener, E., Sandvik, E., Pavot, W., and Fujita, F. 1992a. Extraversion and subjective well-being in a U.S. probability sample. *Journal of Research in Personality*, 26, 205–215.

Diener, E., Sandvik, E., Seidlitz, L., and Diener, M. 1992b. The relationship between income and subjective well-being: Relative or absolute? *Social Indicators Research*, 28, 253–281.

Dweck, C. S. and Leggett, E. L. 1988. A social-cognitive approach to motivation and personality. *Psychological Review*, 95, 256–273.

Eaton, L. G. and Funder, D. C. 2003. The creation and consequences of the social world: An interactional analysis of extraversion. *European Journal of Personality*, 17(5), 375–395.

Emmons, R. A. 1986. Personal strivings: An approach to personality and subjective well-being. *Journal of Personality and Social Psychology*, 51(5), 1058–1068.

Erikson, E. H. 1963. *Childhood and Society* (2nd ed.). New York: Norton.

Erikson, E. H. 1968. *Identity: Youth and Crisis*. New York: Norton.

Fiske, S. T. 2010. *Social Beings: Core Motives in Social Psychology*. New York: Wiley.

Freud, S. 1923/1961. The ego and the id. In J. Strachey (Ed.), *The Standard Edition of the Complete Psychological Works of Sigmund Freud* (Vol. 19). London: Hogarth.

Friedman, H. S. 2000. Long-term relations of personality and health: Dynamics, mechanisms, tropisms. *Journal of Personality*, 68, 1089–1107.

Friedman, H. S. 2008. The multiple linkages of personality and disease. *Brain, Behavior, and Immunity*, 22, 668–675.

Goldberg, L. R. 1993. The structure of phenotypic personality traits. *American Psychologist*, 48, 26–34.

Hammack, P. L. 2008. Narrative and the cultural psychology of identity. *Personality and Social Psychology Review*, 12, 222–247.

Heckhausen, H. 1991. Social bonding: Affiliation motivation and intimacy motivation. In: *Motivation and Action*. Berlin: Springer.

John, O. P. and Srivastava, S. 1999. The big five trait taxonomy: History, measurement, and theoretical perspectives. In L. Pervin and O. P. John (Eds.), *Handbook of Personality: Theory and Research* (2nd ed., pp. 102–138). New York: Guilford Press.

Judge, T. A., Livingston, B. A., and Hurst, C. 2012. Do nice guys—and gals—really finish last? The joint effects of sex and agreeableness on income. *Journal of Personality and Social Psychology*, 102(2), 390–407.

King, L. A. 1995. Wishes, motives, goals, and personal memories: Relations of measures of human motivation. *Journal of Personality*, 63, 985–1007.

Laursen, B., Pulkkinen, L., and Adams, R. 2002. The antecedents and correlates of agreeableness in adulthood. *Developmental Psychology*, 38, 591–603.

LeDoux, J. 1996. *The Emotional Brain*. New York: Touchstone Books.

Lieberman, M. D. and Rosenthal, R. 2001. Why introverts can't always tell who likes them: Multitasking and nonverbal decoding. *Journal of Personality and Social Psychology*, 80(2), 294–310.

Little, B. R. 1983. Personal projects: A rationale and method for investigation. *Environment and Behavior*, 15, 273–309.

Little, B. R. 1999. Personality and motivation: Personal action and the conative evolution. In L. A. Pervin and O. John (Eds.), *Handbook of Personality: Theory and Research* (2nd ed., pp. 501–524). New York: Guilford Press.

Lodi-Smith, J. and Roberts, B. W. 2007. Social investment and personality: A meta-analysis of the relationship of personality traits to investment in work, family, religion, and volunteerism. *Personality and Social Psychology Review*, 11, 68–86.

Lodi-Smith, J. and Roberts, B. W. 2012. Concurrent and prospective relationships between social engagement and personality traits in older adulthood. *Psychology and Aging*, 27, 720–727.

Maslow, A. 1968. *Toward a Psychology of Being* (2nd ed.). New York: Van Nostrand.

Matthews, G. and Gilliland, K. 1999. The personality theories of H. J. Eysenck and J. A. Gray: A comparative review. *Personality and Individual Differences*, 26(4), 583–626.

Mayer, J. D., Faber, M. A., and Xu, X. 2007. Seventy-five years of motivation measures (1930–2005): A descriptive analysis. *Motivation Emotion*, 31, 83–103.

McAdams, D. P. 1985. *Power, Intimacy, and the Life Story: Personological Inquiries into Identity*. New York: Guilford Press.

McAdams, D. P. 2001. The psychology of life stories. *Review of General Psychology*, 5(2), 100–122.

McAdams, D. P. 2015. *The Art and Science of Personality Development*. New York: Guilford Press.

McAdams, D. P. 1995. What do we know when we know a person? *Journal of Personality*, 63, 365–396.

McAdams, D. P. and Manczak, E. 2011. What is a "level" of personality? *Psychological Inquiry*, 22(1), 40–44.

McAdams, D. P. and McLean, K. C. 2013. Narrative identity. *Current Directions in Psychological Science*, 22, 233–238.

McAdams, D. P. and Pals, J. L. 2006. A new Big Five: Fundamental principles for an integrative science of personality. *American Psychologist*, 61, 204–217.

McClelland, D. C. 1961. *The Achieving Society*. New York: D. Van Nostrand.

McClelland, D. C. 1985. *Human Motivation*. Glenview, IL: Scott Foresman

McClelland, D. C., Atkinson, J. W., Clark, R. A., and Lowell, E. L. 1953. *Century Psychology Series. The Achievement Motive*. East Norwalk, CT: Appleton-Century-Crofts.

McCrae, R. R. and Costa, P. T., Jr. 1997. Personality trait structure as a human universal. *American Psychologist*, 52, 509–516.

McCrae, R. R. and Costa, P. T., Jr. 1999. A five-factor theory of personality. In L. Pervin and O. John (Eds.), *Handbook of Personality: Theory and Research* (pp. 139–153). New York: Guilford Press.
McCrae, R. R. and John, O. P. 1992. An introduction to the five-factor model and its applications. *Journal of Personality*, 60(2), 175–215.
McLean, K. C. 2008. Stories of the young and old: Personal continuity and narrative identity. *Developmental Psychology*, 44, 254–264.
McLean, K. C., Pasupathi, M., and Pals, J. L. 2007. Selves creating stories creating selves: A process model of self-development. *Personality and Social Psychology Review*, 11, 262–278.
Mischel, W. 1968. *Personality and Assessment.* New York: Wiley.
Mischel, W. 1973. Toward a cognitive social learning reconceptualization of personality. *Psychological Review*, 80, 252–283.
Murray, H. 2008/1938. *Explorations in Personality.* New York: Oxford University Press.
Noftle, E. E. and Shaver, P. R. 2006. Attachment dimensions and the big five personality traits: Associations and comparative ability to predict relationship quality. *Journal of Research in Personality*, 40(2), 179–208.
Ozer, D. J. and Benet-Martínez, V. 2006. Personality and the prediction of consequential outcomes. *Annual Review of Psychology*, 57(1), 401–421.
Roberts, B. W. and DelVecchio, W. F. 2000. The rank-order consistency of personality from childhood to old age: A quantitative review of longitudinal studies. *Psychological Bulletin*, 126, 3–25.
Roberts, B. W. and Robins, R. W. 2000. Broad dispositions, broad aspirations: The intersection of personality traits and major life goals. *Personality and Social Psychology Bulletin*, 26(10), 1284–1296.
Roberts, B. W. and Robins, R. W. 2004. Person–environment fit and its implications for personality development: A longitudinal study. *Journal of Personality*, 72, 89–110.
Salvendy, G. (Ed.) 2012. *Handbook of Human Factors and Ergonomics* (4th ed.). Hoboken, NJ: Wiley and Sons.
Singer, J. A. 2004. Narrative identity and meaning making across the adult lifespan: An introduction. *Journal of Personality*, 72(3), 437–459.
Singer, J. A. 2005. *Personality and Psychotherapy: Treating the Whole Person.* New York: Guilford Press.
Singer, J. A., Blagov, P., Berry, M., and Oost, K. M. 2013. Self-defining memories, scripts, and the life story: Narrative identity in personality and psychotherapy. *Journal of Personality*, 81, 569–582.
Singer, J. A. and Bluck, S. 2001. New perspectives on autobiographical memory: The integration of narrative processing and autobiographical reasoning. *Review of General Psychology*, 5, 91–99.
Smith, T. W. 2006. Personality as risk and resilience in physical health. *Current Directions in Psychological Science*, 15, 227–231.
Srivastava, S., Angelo, K. M., and Vallereux, S. R. 2008. Extraversion and positive affect: A day reconstruction study of person-environment transactions. *Journal of Research in Personality*, 42(6), 1613–1618.
Wilt, J. and Revelle, W. 2009. Extraversion. In M. R. Leary and R. H. Hoyle (Eds.), *Handbook of Individual Differences in Social Behavior* (pp. 27–45). New York: Guilford Press.
Winston, C. E. 2011. Biography and life story research. In S. Lapan, M. Quartaroli, and F. Riemer (Eds.), *Qualitative Research: An Introduction to Designs and Methods* (pp. 106–136). New Jersey: Jossey-Bass.
Winston-Proctor, C. E. 2018. Toward a model for teaching and learning qualitative inquiry within a core content undergraduate psychology course: Personality psychology as a natural opportunity.*Qualitative Psychology*, 5(2), 243–262.

Winter, D. 1992. Power motivation revisited. In C. Smith (Ed.), *Motivation and Personality: Handbook of Thematic Content Analysis* (pp. 301–310). Cambridge: Cambridge University Press.

Winter, D. G. and Barenbaum, N. B. 1985. Responsibility and the power motive in women and men. *Journal of Personality*, 53, 335–355.

PROBLEMS

These problems are designed to make you think about the essential behavioral science concepts that have been discussed in this chapter. These problems could be used in a number of ways, including as individual thought exercises, group discussion questions, and/or to stimulate interest in new ways of thinking about human personality and behavioral cybersecurity.

2.1 Would it be possible to understand the whole person within the concepts of personality traits, motivation, goals, and narrative identity? Why should we even desire to understand persons to conceptualize and solve cybersecurity problems?

2.2 Identify a recent case of a human hacker in which there is a lot of information about the person from multiple sources (e.g., news reports, case studies, etc.) or select a case within Chapter 4, "Recent Events," and use the case descriptions as a source. Using the dimensions of human personality that have been described in this chapter, try to use clues from the various sources or the Chapter 4 descriptions of recent events about the person to describe his/her personality traits, motivation, and narrative identity.

2.3 What is missing from personality psychologists' model of understanding the whole person as the mission of the field? In other words, what other than personality traits, personality characteristic adaptations, and narrative identity embedded in culture and social context characterize a person's personality?

3 Psychology and Cybersecurity

It is very hard in these times to pick up a newspaper and read about events without at some point encountering a description of some cyberattack, whether it be breaking into computer accounts, downloading personal and private information from large databases, or holding a user hostage by demanding a ransom payment for the user to restore his or her system.

The field of cybersecurity has had enormous development, even going back close to a century, but still the success of the attackers is broadening and growing.

Today's specialists in cybersecurity from a defense perspective consist largely of persons with computer science, mathematics, and engineering backgrounds. Those of us who do research in this area are justifiably proud of the complexity of many of our protocols that have required usually very advanced mathematical techniques to implement and which, when implemented properly, provide very strong means of defense. In many ways, the genesis of this book has been a series of discussions over time by cybersecurity specialists, who on the one hand reflect on the technical achievements in developing these very strong security protocols—but on the other hand recognize that often their strength can be completely obviated by the user who might use their pet's name as their password, thus providing a simple path for an attacker to negate the strong security protocol.

One might argue that the time will come when the human is out of the loop. In other words, at the moment, we might need a human input at some point in a cyberattack scenario, such as by entering a password. At that point, the value of the sophisticated protocol becomes far less important than an understanding of how a human (the defender) fails in providing, for example, a strong password; or how another human (the attacker) is clever or motivated enough to discover the defender's password mechanism.

This argument continues by assuming that if the human is removed—in other words, that whatever is used as a password (or method of authentication) is generated by the computer itself, some automated mechanism, and that the method of breaking it would also be computer generated, the impact of human behavior and behavioral science might be removed from the equation.

The world of cybersecurity may evolve to that point. However, it is difficult to conceive that with the number of individuals participating in the global computer environment now in the billions how all users could live in an environment where all of their modes of identification or authentication would just be in the computer-to-computer domain.

But let's go one step beyond: even given the last assumption, we can also envisage a future where the necessary computer software or robotics to participate in such a dialogue as described to establish communication and therefore elicit meaning of information between two automated systems may also begin to develop behavioral traits. There is a good deal of research going on today on how to ascribe behavioral tendencies to robots, or bots, in other words, through their software design. So perhaps we can see in the future that behavioral scientists may also have to take into consideration the behavioral science of robots.

Indeed, Alan Turing, usually considered the father of computer science, posed this question in the 1940s—we usually refer to the Turing Test, and he described it in an intellectual paper in the journal *Mind* in 1950. Many people saw a form of this discussion in the Academy Award–nominated motion picture *The Imitation Game* in 2015.

Despite the increasing interest and acceleration of cyberattacks worldwide in the present time, to us it has become extremely clear that in order to best understand the overall space of cyberattack/cyberdefense it is insufficient for us only to train in the mathematics- and engineering-related approaches available to the current generation of cybersecurity specialist; we also need the expertise of persons with knowledge in the various behavioral sciences: psychology, behavioral science, neuroscience, and so on. Thus, it is the objective of this book to develop a curriculum suitable for people both with math/computer science/engineering backgrounds and those who are specialists in the various behavioral sciences, to gain an understanding of each scientific approach, and also to develop a core of people with facility in both of these two radically different areas with the capability of analyzing cybersecurity problems.

3.1 WHO SHOULD READ THIS BOOK?

It is our objective that this book should serve multiple audiences. We would like to broaden the field of cybersecurity specialists by providing a path for persons who are trained in mathematics- and engineering-related backgrounds who would also like to understand some principles of behavioral science to be able to bring together these two areas of study. In addition, from another perspective, we would like this book to appeal to behavioral scientists—psychologists, neuroscientists, and others—who are also interested in learning some of the approaches from the more math- and engineering-related fields to similarly become cybersecurity specialists, also with a similar dual approach.

From the perspective of the use of this book as a text, it has been used in the curriculum development and prior offerings as a course at both the upper-division undergraduate and master's level with students who have prior preparation in either mathematics and engineering or behavioral science backgrounds.

This objective of addressing an interdisciplinary audience has provided some challenges from a pedagogical perspective. We have found that we have gotten the deepest involvement from students when an active student learning environment is emphasized: what many might call a form of the "flipped classroom."

We believe that this book can serve several different audiences by selecting several different path through the material, as described in the Preface: a student body that is (a) a mix of computer science/engineering students and behavioral science students, (b) a primarily computer science/engineering class, and (c) a primarily behavioral science class.

REFERENCE

Turing, A. M. 1950. Computing machinery and intelligence. *Mind: A Quarterly Review of Psychology and Philosophy*, LIX(236), 433–460.

PROBLEMS

3.1 Find a job website that gives a description of a cybersecurity expert. What is that description, and what is the source?

3.2 This chapter describes one worldview where we humans would no longer have to remember passwords in our computer usage.
   a. When in the future might this occur?
   b. What problems could this solve?
   c. What problems could this create?

3.3 Construct a separate example. Allow a classmate five guesses. How many of your classmates will find the password?

3.4 Assess the published movie reviews of *The Imitation Game*.

3.5 Does this film adequately describe the Turing Test?

3.6 Define a "flipped classroom."

3.7 Identify a university catalog where (a) a computer science major is required to take a psychology course and (b) a psychology major is required to take a computer science course.

4 Recent Events

The history of research in cybersecurity dates back to the 1970s and even before, but for most people, what was known at the time affected only a very small number of people in the world of computing. As has been described above, the first general awareness of external attacks occurred only in the 1980s with, for example, the Morris Internet worm of 1988.

4.1 MORRIS WORM

On November 2 of that year, Robert Morris, then a graduate student in computer science at Cornell, created the worm in question and launched it on the Internet. In UNIX systems of the time, the applications sendmail and finger had weaknesses that allowed the worm to enter and then generate copies of itself. This resulted in the major damage caused by this work, as it would make copies of itself until memory was exhausted, causing the system to shut down. Furthermore, with UNIX vulnerabilities, the worm could move from one machine to the other, and it was estimated that it eventually infected about 2000 computers within 15 hours. The fix for each machine consisted only of deleting the many copies of the worm; no data was modified or removed. The time to produce a fix varied between a few hours to days. The U.S. Government Accountability Office estimated the cost of the damage in the range of $100,000–$10 million—obviously not a very accurate assessment (Stoll, 1989).

Robert Morris was tried and convicted under the Computer Fraud and Abuse Act, and was sentenced to 3 years probation and 400 hours of community service, and fined $10,050. He was subsequently hired as a professor of computer science at the Massachusetts Institute of Technology, where he continues to teach and research to this day.

The proliferation of access to computers to the general public, both in North America and worldwide, only began in the 1990s. Consequently, for most of us, hearing of cyberattacks only began in that period.

But now, in the twenty-first century, the percentage of the global population with access to the Internet has increased exponentially, and consequently the number of targets for cyberattackers and, undoubtedly, the number of attackers have also increased.

Not only that, but the potential for exploitation of many different types of target has also increased. Therefore, we now have many examples of exploits that result in widespread theft of critical user information such as Social Security numbers, addresses, telephone numbers, and even credit card information, for example, the following "OPM hack."

4.2 THE OFFICE OF PERSONNEL MANAGEMENT HACK

The Office of Personnel Management (OPM) is a little-known but important component of the United States government. It is responsible for "recruiting, retaining and honoring the world-class force to serve the American people." In other words, OPM handles a good deal of the process of record keeping for employees of the United States government; in particular, it also manages security clearances for many federal employees.

Two major data breaches occurred in 2014 and 2015 that obtained the records of approximately 4 million U.S. government employees, with the estimate of the number of stolen records as approximately 21.5 million. Information obtained included Social Security numbers, names, dates and places of birth, and addresses. Also stolen in this attack were 5.6 million sets of fingerprints. As we will see later in this book, obtaining much of this information can often be done without data breaches (see the Hack Lab on Sweeney's Research). However, gaining this information legally usually requires finding this data one record at a time rather than over 20 million in one fell swoop (OPM, 2018).

There were two attacks at the OPM. It is not known when the first attack (named X1 by the Department of Homeland Security) occurred, but it was discovered on March 20, 2014, the second attack (later called X2) occurred on May 7, 2014, but was not discovered until April 15, 2015.

In the second attack, the hackers posed as employees of a subcontractor of the government, KeyPoint Government Solutions. As a consequence of these two attacks, both the director and chief information officer of OPM resigned.

It has been suggested that the malicious software was so highly developed that the attackers may have been Chinese government officials.

4.3 WIKILEAKS

Attacks on many organizations use techniques such as distributed denial of service (DDoS) designed primarily to cripple an organization's website and thus prevent the organization from doing business, even if only for a relatively short period of time.

Perhaps the first time there was a widespread public awareness of DDoS attacks occurred around November 28, 2010. The web site wikileaks.org published several thousand classified U.S. government documents and announced it had 250,000 more. This caused an immediate "cyberwar," with opponents of WikiLeaks attempting to crash the WikiLeaks website, while WikiLeaks supporters, particularly a loosely affiliated group called Anonymous, were able to crash numerous sites, including MasterCard and PayPal.

4.3.1 What's a WikiLeaks?

In 2007, WikiLeaks was founded as an organization and a website whose mission was to:

Bring important news and information to the public
Provide an innovative, secure, and anonymous way for sources to leak information to journalists
Publish original source material alongside news stories so readers and historians alike could see evidence of the truth (WikiLeaks, 2018)

"The broader principles on which our work is based are the defense of freedom of speech and media publishing, the improvement of our common historical record and the support of the rights of all people to create new history. We derive these principles from the Universal Declaration of Human Rights."

4.3.2 A Noble Objective

Those objectives are laudable, and if we were only talking about

The looting of Kenya under President Daniel Arap Moi—$3,000,000,000 of presidential corruption exposed

Stasi (former East German Secret Police), still in charge of Stasi files, suppressed a 2007 investigation into infiltration of former Stasi into the Stasi Files Commission

Exposure of a press gagging order from the Turks and Caicos Islands

(all of these discoveries contributed to several international human rights awards for WikiLeaks). Well, we'd be pretty happy about WikiLeaks, and might even donate a little money. Alas, around November 28, 2010, WikiLeaks released several thousand classified U.S. government documents and announced they had 250,000 more. (It was later determined that the documents had been provided by U.S. Army Private Chelsea Manning.)

There was immediately a firestorm of protest from all quarters in the United States, and WikiLeaks founder Julian Assange's head was demanded on a plate!

4.3.3 TREASON!!!

Such notables as Representative Peter King, ranking Republican on the House Homeland Security Committee, and commentator Bill O'Reilly all called for Assange's prosecution under a charge of treason. There was only one small problem here: by law, treason in the United States can only be committed by a U.S. citizen—and Assange is Australian.

Bob Beckel of Fox News (and a onetime campaign manager for Democratic Vice President Walter Mondale) had also called Assange treasonous, but he has advocated not bothering with a trial and just killing him outright.

But the immediate response from the cybercommunity was that two of the main sources of receipt of donations to WikiLeaks, MasterCard and PayPal, announced that they would no longer accept donations destined as contributions to WikiLeaks—in fact, donations were the primary source of revenue for WikiLeaks.

Subsequently, organizations supportive of the objectives of WikiLeaks, notably an online group called Anonymous, decided to launch DDoS attacks on both MasterCard and PayPal and were successful in bringing them down for several days, undoubtedly costing both those organizations significant loss of revenue. In retaliation, in effect perhaps the first example of cyberwarfare, other groups sympathetic to MasterCard and PayPal and in opposition to WikiLeaks' release of the classified documents made

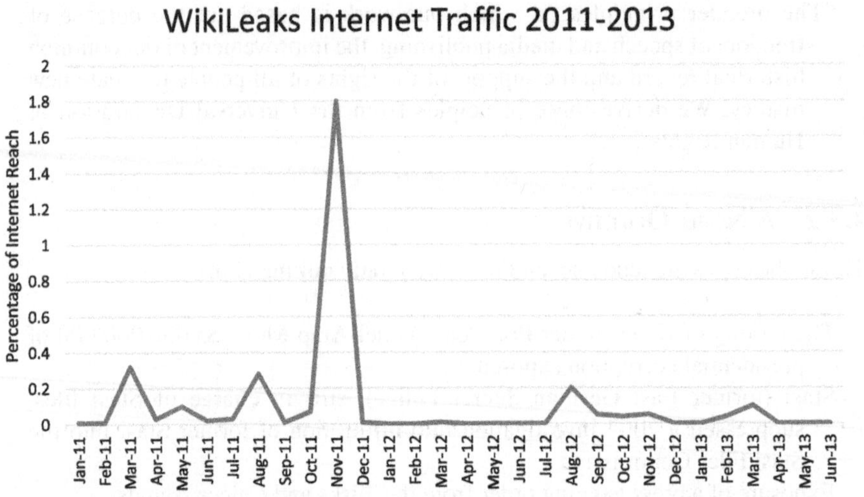

FIGURE 4.1 Traffic on the WikiLeaks website showing attack level after Manning documents leaked.

a similar attempt to take down the WikiLeaks website, but WikiLeaks, in anticipation of such an action, had strengthened their defense against such attacks so that they were never successful in shutting down the WikiLeaks site.

At the time, the online traffic monitoring company Alexa (2018) made public data on the traffic of many websites around the world, and the following chart of Alexa data showed the strength of the attack on WikiLeaks, even though it did not succeed in causing WikiLeaks to shut down (Figure 4.1).

A new type of attack, ransomware, made its first known appearance in 1989. These are attacks designed to disable a person's computer and information until a sum of money is paid to the perpetrator.

4.4 EARLY RANSOMWARE

The first known case of ransomware is about as bizarre a story as in the entire annals of cybersecurity. It dates back to 1989 (pre-Web), and it was launched—on floppy disks—by an evolutionary biologist and Harvard PhD, Joseph Popp. Popp was an AIDS researcher and a part-time consultant for the World Health Organization in Kenya. His ransomware involved copying a virus called the AIDS Trojan that encrypted a victim's files after a certain number of reboots of their system (Simone, 2015). He mailed this virus to approximately 20,000 recipients on floppy disks. In order to respond to the demand for ransom, $189 was to be paid to a Panamanian post office box; then the victim would receive decryption software. The attack was largely unsuccessful, since upon analysis of the code, decryption tools were made freely available. Popp was subsequently arrested in England but was subsequently declared mentally incompetent to stand trial. Later he returned to the United States and at one point created the Joseph L. Popp, Jr., Butterfly Conservatory in Oneonta, New York, which is an existing attraction in that city, only a few miles away from a perhaps more famous institution, the Cooperstown, New York, Baseball Hall of Fame. He built the Conservatory with his daughter, but he passed away shortly before it was inaugurated in 2006.

Ransomware resurfaced in 2013 with CryptoLocker, which used Bitcoin to collect ransom money. In December of that year, ZDNet estimated based on Bitcoin transaction information that the operators of CryptoLocker claimed about $27 million was procured from infected users.

In a nonencrypting ransomware variation, in 2010, WinLock restricted access to a system by displaying pornographic images and asked users to send $10 to receive a code to unlock the machines. This scheme reportedly collected $16 million, although a user with some knowledge could have easily defeated the scheme.

Another type of malware, "phishing," attempts to deceive a user into giving up sensitive information, most often by merely opening a "phishing" email, or even by responding to such an email by returning sensitive information.

4.5 PHISHING

Phishing attacks are a form of illicit software designed primarily to obtain information to benefit the "phisher" from an unsuspecting person or account.

These attacks might arise from any source the user contacts for information, and many might occur from opening an email supposedly from some trusted source.

The purpose for the attack might be urging the recipient to open an attachment. Many users might not realize that a Word, Excel, or PowerPoint document may contain code (called a macro) that may then infect his or her system.

Another approach in a phishing attack might be to encourage the recipient to follow a link that purports to require the user to enter an account name, password, or other personal information, which then is transmitted to the creator of the phishing attack. In such a case, the personal information transmitted may be used by the phishing perpetrator in order to gain other resources of the victim.

One example from 2016 involves an email allegedly from the PayPal Corporation. Figure 4.2 shows a screenshot of the attack itself. In this case, the objective of the attack is to fool the recipient into believing that this is a legitimate email from PayPal, asking for "help resolving an issue with your PayPal account," consequently passing on the login information to the phishing attacker.

A prudent user would look carefully at the email address and service, paypal@outlook.com, and realize that it was a bogus email address (Figure 4.2).

On a larger scale, the use of malicious software and sometimes hardware has been used in order to damage or destroy critical infrastructure.

4.6 STUXNET

The year 2010 produced a "game changer." For perhaps the first time, a malicious hardware and software attack, called Stuxnet, infected nuclear facilities in Iran. One critical difference here was that previous malware was *always* produced by individuals or small groups, sought random targets, was easily disabled when identified, and caused relatively minimal damage.

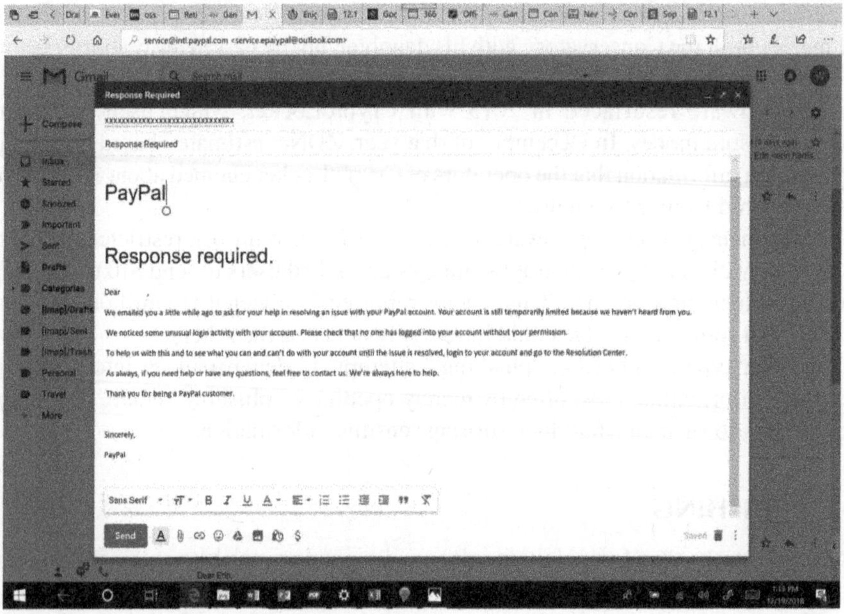

FIGURE 4.2 A phishing attack using PayPal.

Stuxnet was discovered by security researchers. It was determined to be a highly sophisticated worm that spread via Windows and targeted Siemens software and equipment (it was known that the Siemens SCADA—Supervisory Control and Data Acquisition system—was the software managing the Iranian nuclear power development systems).

This was the first malware discovered that subverted industrial systems and the first to include a programmable logic controller (PLC) rootkit.

4.6.1 What Does Stuxnet Do?

Initially, it just tries to access all computers possible (like previous worms). But then, if the attacked computer is not a Siemens SCADA system, then no harm done. If it is a SCADA system, then it infects PLCs by corrupting the Step-7 software application that reprograms the devices.

4.6.2 So What Happened?

Different versions of Stuxnet infected five Iranian organizations, presumably related to the uranium enrichment infrastructure. The Symantec Corporation reported in August 2010 that almost 60% of Stuxnet-infected computers worldwide were in Iran.

4.6.3 Result

The Iranian nuclear program was damaged by Stuxnet, as the infected control system created a change in temperature in the core, thus destroying the equipment.

Kaspersky Labs concluded the attacks "could only have been conducted with nation-state support."

Stuxnet does little harm to computers and networks that do not meet specific configuration requirements; "The attackers took great care to make sure that only their designated targets were hit… It was a marksman's job." (Sanger, 2012). The worm became inert if Siemens software was not found.

4.6.4 And Who Did It?

Symantec estimates that the developing group had to have 5 to 30 people, and the worm took 6 months to prepare.

Bruce Schneier stated at the time that "we can now conclusively link Stuxnet to the centrifuge structure at the Natanz nuclear enrichment lab in Iran" (Schneier, 2012).

Ralph Langner, who first identified what Stuxnet did: "My opinion is that [Israel] is involved but that the leading force is not Israel … it is the cybersuperpower—there is only one, and that's the United States." (Broad et al., 2011; Langner, 2011).

The Iranian government admitted in November 2010 that a virus had caused problems with the controller handling the centrifuges at Natanz.

4.6.5 No Person or Country Originally Admitted Ownership of Stuxnet

Later, in 2013, U.S. General James Cartwright revealed details about Stuxnet. See: http://thecable.foreignpolicy.com/posts/2013/09/24/obamas_favorite_general_stripped_of_his_security_clearance (Lubold, 2013).

See also (Sanger, 2012): http://www.nytimes.com/2012/06/01/world/middleeast/obama-ordered-wave-of-cyberattacks-against-iran.html?pagewanted=all.

Later reports indicated that Stuxnet was a joint effort of the U.S. National Security Agency and the comparable Israeli agency, the Mossad.

The "success" of Stuxnet spawned several later variations.

4.6.6 Duqu

On September 1, 2011, a new worm was found, thought to be related to Stuxnet. The Laboratory of Cryptography and System Security (CrySyS) of the Budapest University of Technology and Economics analyzed the malware, naming the threat Duqu. Symantec, based on this report, continued the analysis of the threat, calling it "nearly identical to Stuxnet, but with a completely different purpose," and published a detailed technical paper. The main component used in Duqu is designed to capture information such as keystrokes and system information. The exfiltrated data may be used to enable a future Stuxnet-like attack.

4.6.7 Flame

In May 2012, the new malware "Flame" was found, thought to be related to Stuxnet. Researchers named the program Flame after the name of one of its modules. After analyzing the code of Flame, Kaspersky said that there was a strong relationship between Flame and Stuxnet. An early version of Stuxnet contained code to propagate infections via Universal Serial Bus (USB) drives that is nearly identical to a Flame module that exploits the same zero-day vulnerability.

4.6.8 Gauss

In August 2012, Goodin reported, "Nation-sponsored malware with Stuxnet ties has mystery warhead." Adding to the intrigue, the Gauss Trojan also targeted Middle East banks and PayPal (Goodin, 2012). Gauss, as Kaspersky Lab researchers have dubbed the malware, was devised by the same "factory" or "factories" responsible for the Stuxnet worm used to disrupt Iran's nuclear program, as well as the Flame and Duqu Trojans.

In addition to the examples mentioned briefly above, all of these levels of attacks are on the increase, and despite our best efforts in the cybersecurity community, the factor that is most difficult to control is not our software and hardware to provide defense—but instead the human frailties of users.

Just taking 2017 and 2018 for examples, we have the following.

4.7 PRESIDENTIAL ELECTION

Attacks that succeeded in obtaining significant email traffic from the internal communications of the United States Democratic Party and its presidential candidate, Hillary Clinton, in all likelihood had a significant impact on the outcome of the election. In particular, Clinton's campaign chairman John Podesta had many of his emails stolen, as he was tricked by a phishing attack to allow an outsider into his email (Figure 4.3).

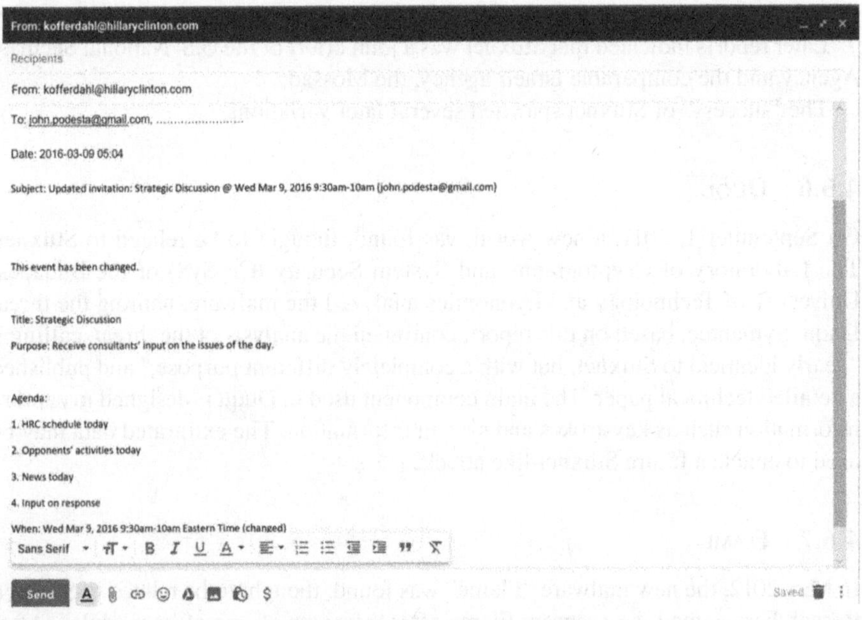
From: kofferdahl@hillaryclinton.com

Recipients

From: kofferdahl@hillaryclinton.com

To: john.podesta@gmail.com, ...

Date: 2016-03-09 05:04

Subject: Updated invitation: Strategic Discussion @ Wed Mar 9, 2016 9:30am-10am (john.podesta@gmail.com)

This event has been changed.

Title: Strategic Discussion

Purpose: Receive consultants' input on the issues of the day.

Agenda:

1. HRC schedule today
2. Opponents' activities today
3. News today
4. Input on response

When: Wed Mar 9, 2016 9:30am-10am Eastern Time (changed)

Sans Serif

Send

Saved

FIGURE 4.3 Podesta email hacked and sent to WikiLeaks.

Early in 2018, ransomware attacks known as WannaCry and Petya were discovered; the former apparently put close to 100,000 computers up for ransom, including computing systems in many hospitals in the United Kingdom, and Petya disabled many industries in Ukraine.

4.8 WANNACRY AND PETYA

4.8.1 A Case Study: Two Major Cyberattacks of 2017: Alike or Different?

On May 13, 2017, many organizations were thrown for a loop with a ransomware attack that became known as WannaCry (Figure 4.4). Throughout England, for example, many hospitals had to shut down. The impact, on an estimated 300,000 computers worldwide, was that users' displays were frozen, with a message indicating that the user's files had been encrypted, demanding payment of $300 (or the equivalent in Bitcoin) in order for the files to be decrypted.

On or about June 25, after the initial reactions and defenses to WannaCry had been developed, a new attack emerged, generally called Petya (Figure 4.5). Petya is also ransomware, and has many of the same objectives as WannaCry, but clearly there were some differences.

In order to understand the sequence of events during both WannaCry and Petya, a number of accounts of the events should be read. Wikipedia (2017a,b) is a key to the reading for both cyberattacks.

These cases can be studied from different levels for students in different courses.

FIGURE 4.4 WannaCry screen.

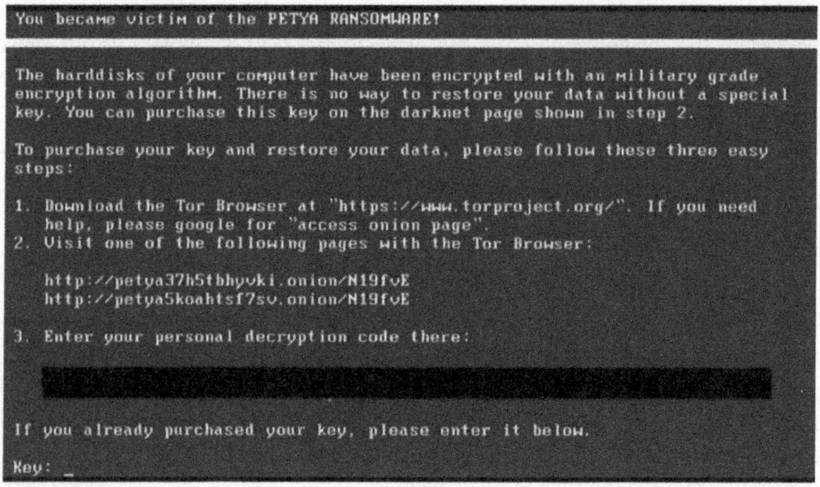

FIGURE 4.5 Petya screen.

4.8.2 UNDERSTANDING THE WANNACRY ATTACK (5/13/17) PLAY-BY-PLAY

First, a chronological description of events leading up to the current attack that made it possible:

1. Windows, throughout its history, first installed updates (for security or otherwise) by the user specifically downloading and installing various patches.
2. By the time Windows evolved through generations (e.g., Windows 7 and XP), the updates could be installed automatically, but the user had to turn on the feature for automatic downloads. Many people didn't.
3. With the current Windows 10, the downloads are automatically installed every time you power off the computer.
4. Some time ago, the National Security Agency developed extensive software to hack into Windows systems through a weakness in the server message block protocol. This was called EternalBlue.
5. Perhaps as early as 2014, someone, or perhaps several individuals, were able to acquire this EternalBlue package.
6. Through whatever mechanism, a hacker group advertising itself as the Shadow Brokers began to try to sell the EternalBlue package through the underground.
7. The language they used in advertising this product was bizarre, to say the least.
8. In March 2017, Microsoft realized there was a weakness in the server message block protocol and issued a way of defeating the type of attack that could use that weakness.
9. Which, of course, was the weakness that EternalBlue knew how to exploit.
10. Not only did Microsoft provide this fix automatically for Windows 10 machines, it also provided a fix for the earlier versions—but because Microsoft is no longer supporting systems like Windows 7 or XP, the updates would not be automatically installed.

11. Anyone or any group acquiring this package would have solved half the problem of launching a ransomware attack. They would have the means of gaining entry into many Windows systems that did not have the automatic update feature.
12. The final step was for a hacker group to create a form of ransomware that would enter systems using its EternalBlue entry mechanism, then launch the WannaCry software that would freeze the user screen, encrypt some or all files, and then demand a ransom.
13. This attack was launched Friday, May 13, 2017, and there are indications are that WannaCry has infected 300,000 systems.
14. Windows systems that had installed the Microsoft patch of 2 months before (whether automatically as with a Windows 10 system or on following Microsoft's instructions) could defeat the EternalBlue/WannaCry intrusion.
15. The ransom requested initially was $300, using Bitcoin.
16. A few days after the attack, it had been tracked that only about 900 of the 100,000 systems infected before the weekend had paid the $300.
17. Indications were subsequently that the hacker group was a somewhat well-known group called Lazarus.
18. Some writers believe that Lazarus is either the North Korean government or an organization in the employ of the North Korean government.
19. Lazarus was also mentioned 3 years ago during the so-called Sony Pictures hack.
20. At the height of the WannaCry attack, a researcher in Scotland discovered that as one part of the attack, the WannaCry software checks for the existence of a bogus website called "iuqerfsodp9ifjaposdfjhgosurijfaewrwergwea.com." In the absence of such a website, WannaCry continues its ransomware demand; otherwise, it shuts down. So the enterprising Scot bought a website under that name for about $10, and subsequently infestations of WannaCry stopped because it found the existence of that site.
21. A number of researchers have tried to track the actual impact of the ransomware demand. It is not clear that those who made the Bitcoin payment (a) either had the payment reach its intended destination, nor (b) if any of their files were restored.
22. With respect to the EternalBlue theft from the National Security Agency, two NSA employees have been charged. Nghia Pho has been sentenced to 5½ years in federal prison for Willful Retention of Classified Information. Harold T. Martin III, who was also a graduate student in computer science at the University of Maryland Baltimore County, pleaded guilty to one charge of Willful Retention of National Defense Information.

4.8.3 Understanding the Petya Attack (6/27/17) Play-by-Play

First, a chronological description of events leading up to the current attack and what made it possible:

1. On June 27, 2017, beginning in Ukraine (1 month after WannaCry), a number of banks, energy companies, and utilities had multiple computers attacked and frozen for ransom.

2. Similar results occurred with the Danish shipping company Maersk.
3. Other companies brought their systems down as a preventative response.
4. A system under attack displays a message with a specific email address for payment of the ransom, which is $300, as with WannaCry.
5. Soon, that email address was shut down.
6. It is reported that nine people paid the ransom.
7. The Kiev metro system, Ukrainian gas stations, and the country's deputy prime minister were hit.
8. It was observed that Petya does not try to encrypt individual files, but encrypts the master file table.
9. It was determined that the Petya developers used the EternalBlue package to exploit the Windows vulnerability.
10. It was also noted that the initial Petya attack was via email in a malicious link from an unknown address.
11. Petya ransomware continued to spread rapidly across the globe, impacting multiple corporations and utilities, and it was revealed that the attacker's email address needed to pay the ransom had been shut down, eliminating that possibility for any victim.
12. The attack looked as if it may have started in Ukraine, where banks, energy companies, an airport, and its metro network were affected, according to a *Forbes* report and additional sources. Outside Ukraine, in addition to Maersk, infections also apparently hit British advertiser WPP and the Russian oil industry company Rosnoft.
13. "We can confirm that Maersk IT systems are down across multiple sites and business units due to a cyberattack," read a notification on Maersk's home page. "We continue to assess the situation. The safety of our employees, our operations and customer's business are our top priority. We will update when we have more information."
14. The Chernobyl nuclear power plant in the Ukraine was also hit. The plant was destroyed in a meltdown in 1986 but is still being decommissioned.
15. The German email provider Posteo reported it shut down the email address that the Petya attackers set up to receive ransom payments.
16. "This email address [wowsmith123456@posteo.net] is displayed in Petya's ransom note as the only way to contact the Petya author. Victims have to pay the ransom and send an email with their Bitcoin wallet ID and infection key to the author," *Bleeping Computer* reported, which meant there was no longer any method in place for those with locked files to have them decrypted.
17. Reports from a number of security companies alleged that the ransomware locked up systems globally, including pieces of critical infrastructure and government bodies in Ukraine. Kiev's Boryspil airline said it could cause flights to be delayed, the BBC reported.

4.8.4 WannaCry and Petya: The Same or Not?

There were numerous reasons on the surface to consider that WannaCry and Petya had come from the same source, which might lead to a clue as to the motivation of the perpetrator or perpetrators. Further analysis here might suggest using the profiling methods from Chapter 5 to determine the attackers. Consider these similarities:

WannaCry	Petya
Begins May 13	Begins June 25
Freezes screen	Freezes screen
Demands a Bitcoin payment ($300)	Demands a Bitcoin payment ($300)
Encrypts many files; payment gets decryption key (it says)	Encrypts master file table; payment gets key (it says)
Spread by Eternal Blue suite	Spread by email from anonymous address
Shuts down UK hospitals	Shuts down Ukraine energy companies
Documented 150 people paid	Documented 9 people paid

After this analysis, however, the differences begin to appear. One primary conclusion is that Petya is only masquerading as ransomware. After all, it seems that it only attracted 9 paydays. On further analysis, it seemed that the real purpose of Petya was to be a form of a DDoS.

Nick Bilogorskiy, Cyphort's senior director of threat operations, issued an early breakdown of how the ransomware operated and how it differed from WannaCry. "This is what Petya is, an older ransomware family that has been given a new life by embedding a way to self-replicate over SMB using Eternal Blue exploit," he said, adding that so far nine people had forked over the $300 ransom. Cyphort discovered there were a few differences from WannaCry, namely:

- Petya's initial distribution method was over email, in a malicious link sent from an unknown address.
- Once executed, Petya did not try to encrypt individual files, but encrypted the master file table.
- It had a fake Microsoft digital signature appended, copied from Sysinternals.
- It looks like this new variant could also spread laterally using Windows Management Instrumentation (WMI).
- Some payloads included a variant of Loki Bot, which is a banking Trojan that extracts usernames and passwords from compromised computers, in addition to the ransomware.

4.9 YU PINGAN

Some malicious attacks have a long lifespan. The attacks against the OPM described above only became unraveled in August 2017, when a sealed indictment was filed in the Southern District of California against the Chinese national named Yu Pingan (also known as Gold Sign) on a charge of Conspiracy Computer Hacking.

The charge alleges that Mr. Pingan provided malicious software, in particular a software tool known as "Sakula," to four companies, two in California, one in Massachusetts, and one in Arizona. This software was capable of downloading to

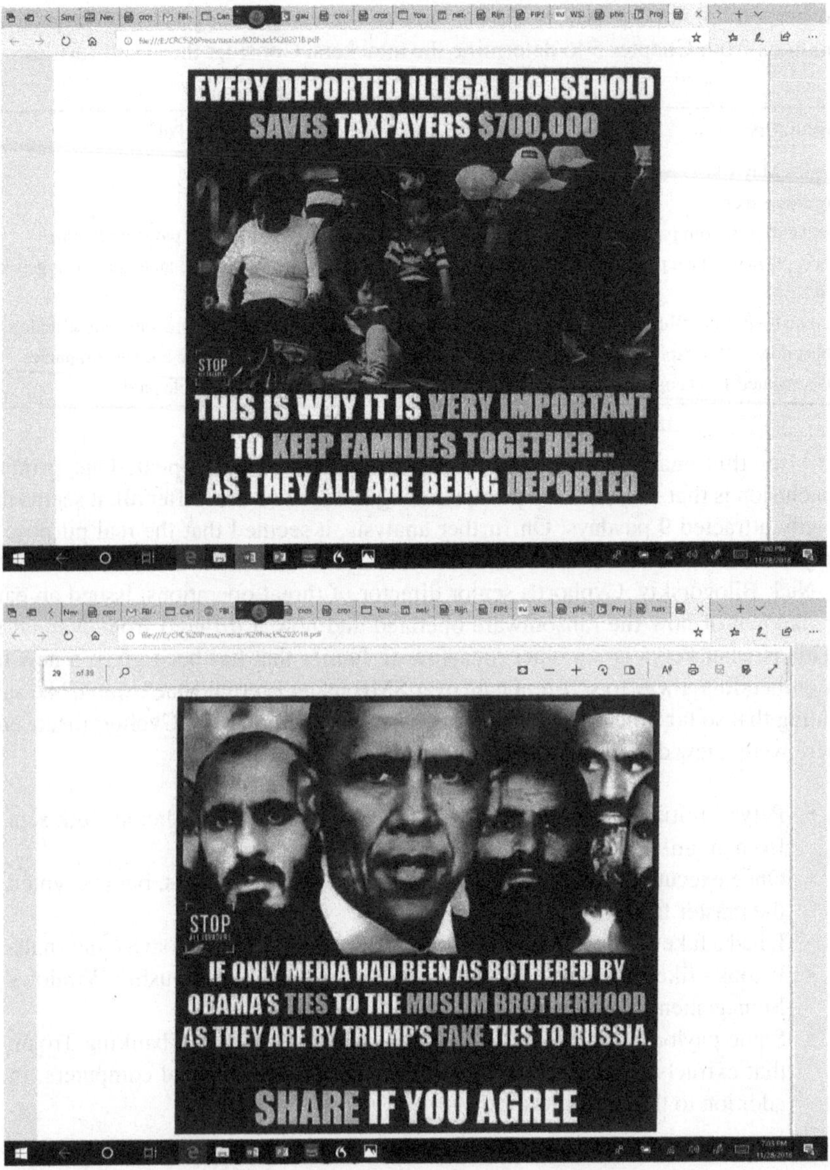

FIGURE 4.6 Facebook posts listed in indictment of Ms. Khusyaynova.

victims' computers without authorization. Sakula was then alleged to have been necessary for large-scale breaches, including the OPM attack, as well as the health insurer Anthem.

Mr. Pingan was at the time a 36-year-old resident of Shanghai, China. The complaint was filed and he was subsequently arrested because he flew to the United States to attend a conference and thus was arrested on U.S. soil.

To quote from the complaint: "in September 2012, malicious files were installed on company sees Web server as part of a watering hole attack that, between September 18, 2012 and September 19, 2012, distributed malicious code to 147 unique US-based IP addresses, using a zero day exploit now known as CEU-2012-4969. Between May 2012 January 2013, company sees Web server hosted no less than five variants of Internet Explorer zero-day exploits" (*US v. Pingan Yu*, 2018).

4.10 ELENA ALEKSEEVNA KHUSYAYNOVA

As an indication that the state-directed attack on the U.S. electoral system continues beyond what was reported above regarding the 2016 U.S. presidential campaign, on September 28, 2018, a criminal complaint was filed in the Eastern District of Virginia of Conspiracy to Defraud the United States against Elena Alekseevna Khusyaynova (DoJ, 2018). She was a resident of St. Petersburg, Russia, and had been the chief accountant in project Lakhta, which has been supported by Yevgeniy Nick Tarter Mitch Prigozhin, closely related to Pres. Vladimir Putin and often referred to as "Putin's chef." She had moved to the Washington, DC, area and was accepted into a graduate program at the George Washington University.

As part of her efforts in project Lakhta, she is alleged to have conducted "information warfare against the United States of America, through fictitious US persons on social media platforms and other Internet-based media." This conspiracy launched many fictitious campaigns designed to sow discord among persons of various political persuasions in the United States and to further the political divisions that now exist. Several examples of the Facebook posts attributed to this effort include (Figure 4.6).

REFERENCES

Alexa Internet, Inc. 2018. http://www.alexa.com

Broad, W. J., Markoff, J., and Sanger, D. E. 2011. *Israeli test on worm called crucial in Iran nuclear delay.* New York Times, January 16. https://www.nytimes.com/2011/01/16/world/middleeast/16stuxnet.html

Department of Justice. 2018. *Office of Public Affairs.* Russian National Charged with Interfering in U.S. Political System, October 19. https://www.justice.gov/opa/pr/russian-national-charged-interfering-us-political-system

Goodin, D. 2012. *Nation-Sponsored Malware with Stuxnet Ties Has Mystery Warhead.* ArsTechnica, August 9. https://arstechnica.com/information-technology/2012/08/nation-sponsored-malware-has-mystery-warhead/

Langner, R. 2011. *Cracking Stuxnet, a 21st Century Weapon.* TED2011, March. https://www.ted.com/talks/ralph_langner_cracking_stuxnet_a_21st_century_cyberweapon?language=en

Lubold, G. 2013. *Obama's Favorite General Stripped of His Security Clearance.* Foreign Policy, September 24. http://thecable.foreignpolicy.com/posts/2013/09/24/obamas_favorite_general_stripped_of_his_security_clearance

Office of Personnel Management. 2018. https://www.opm.gov

Sanger, D. E. 2012. *Obama order sped up wave of cyberattacks against Iran.* New York Times, June 1. https://www.nytimes.com/2012/06/01/world/middleeast/obama-ordered-wave-of-cyberattacks-against-iran.html?pagewanted=all

Schneier, B. 2012. Another piece of the Stuxnet puzzle. *Schneier on Security*, February 23. https://www.schneier.com/blog/archives/2012/02/another_piece_o.html
Simone, A. 2015. *The Strange History of Ransomware*. Medium.com. *Unhackable: An Intel Security publication on Medium*. "Practically Unhackable," March. https://medium.com/un-hackable/the-bizarre-pre-internet-history-of-ransomware-bb480a652b4b
Stoll, C. 1989. *The Cuckoo's Egg*. New York: Doubleday, 336 p.
US v. Pingan Yu. 2018. Case No. 17CR2869-BTM. *UNITED STATES OF AMERICA, Plaintiff, v. PINGAN YU, a.k.a. "GoldSun," Defendant*. United States District Court, S.D. California. August 9.
WikiLeaks. 2018. https://wilileaks.org
Wikipedia. 2017a. *Petya Ransomware Attack*. Wikipedia May. https://en.wikipedia.org/wiki/Petya_(malware)
Wikipedia. 2017b. *WannaCry Ransomware Attack*. Wikipedia May. https://en.wikipedia.org/wiki/WannaCry_ransomware_attack

PROBLEMS

4.1 Estimate the number of computers infected by the Morris worm.
4.2 Discuss the General Accounting Office estimates of the cost of the Morris worm. Find alternative estimates.
4.3 What courses is Robert Morris offering at MIT this year?
4.4 Where is the OPM in the United States government administration?
4.5 What organization or organizations have given awards to WikiLeaks for humanitarian efforts?
4.6 Find an organization for which one can make donations through PayPal.
4.7 Identify a ransomware attack and an estimate of the amount of funds paid to the perpetrator or perpetrators.
4.8 What is the difference between phishing and spear-phishing?
4.9 How many emails related to the 2016 presidential election were estimated to be released by WikiLeaks?
4.10 You can find on the WikiLeaks website the text of the emails to and from John Podesta during the 2016 presidential election campaign. Select 10 of these for further analysis. Rank them in terms of their potential damage.
4.11 At what point in the investigation of WannaCry and Petya could you determine that they had different objectives?
4.12 What is the difference between Willful Retention of Classified Information and Willful Retention of National Defense Information?

5 Profiling

There is a technique in law enforcement called "profiling," which has been used over many years in order to determine if a given criminal behavior leads to a mechanism for defining a category of suspects.

Although this approach has been successful in many instances, it has also led to widespread abuses. First, let's consider one of the major types of abuse. The expression "driving while black" has evolved from the all-too-common practice of police officers stopping an automobile driven by an African-American who might be driving in a predominantly white neighborhood. As eminent a person as former President Barack Obama has reported that this has happened to him on occasion (Soffer, 2016).

Until fairly recently, the sheriff of the largest jurisdiction adjacent to the city of New Orleans had publicly taken the position that in the largely white neighborhoods of his jurisdiction, his officers would routinely stop cars in that neighborhood driven by African-Americans.

Thus, in the realm of cybersecurity, it is important to recognize that criteria that might be used to develop profiling approaches need to be sensitive in attributing certain types of behavior based on race, ethnicity, or other identifiable traits of an individual, rather than attributing types of behavior based on actions rather than individual traits.

To consider an example, let us take the Turing Test model and adapt it to the following situation.

In the first setting, we have an environment where the "profiler" sits at a computer, and in another part of the room, the "subject" (who is visible to the profiler) is also at another computer, and both enter a dialogue where the profiler will attempt to determine some behavioral pattern of the subject. In this dialogue, the profiler notices over time that every time the subject types a personal pronoun in the third person, the subject types "she." The profiler, looking across the room, perceives that the subject is female, and deduces that the reason the subject uses "she" rather than "he" is the fact that the subject is indeed female.

As a second setting for the same experiment, in this case as in the Turing model, the subject is in a different room not visible to the profiler (Turing, 1950). Once again, in the dialogue, the profiler may perceive that the use of the third person pronoun always being "she" would lead to the conclusion that the subject is indeed female.

Of course, this may or may not be the case, since in the second instance, the subject might have made a conscious decision—perhaps to deceive the profiler—to always use "she" rather than "he." In any case, the profiler in the first setting might decide on the gender of the subject merely by the visual appearance rather than by analysis of the message that has been typed. This could not be the case for any conclusion deriving from the second setting of the experiment.

5.1 PROFILING IN THE CYBERSECURITY CONTEXT

There has been a growing body of instances of cyberattacks where there is a great need to try to isolate one or several potential perpetrators of the attack.

Throughout this book, you will see a number of case studies of both actual and fictional cyberattacks, and the profiling techniques described here may be applicable in determining potential suspects. To introduce the subject, however, we will use one well-known series of incidents that we might describe as the "Sony Pictures hack."

5.2 SONY PICTURES HACK

In October 2015, word leaked out that Sony had under development a film titled *The Interview* (Sony, 2018). What was known was that the storyline for this film was that the United States government wanted to employ journalists to travel to North Korea to interview the president of that country—and, under the cover of the interview, to assassinate the president.

As word of this plot leaked out, the North Koreans—understandably furious—threatened reprisals for Sony and for the United States should this film be released.

On November 24, 2014, a hacker group that identified itself by the name Guardians of Peace (GOP) leaked a release of confidential data from the film studio Sony Pictures. The data included personal information about Sony Pictures employees and their families, emails between employees, information about executive salaries at the company, copies of then-unreleased Sony films, and other information. The perpetrators then employed a variant of the Shamoon wiper malware to erase Sony's computer infrastructure (Alvarez, 2014).

Also in November 2014, the GOP group demanded that Sony pull *The Interview*, a comedy about a plot to assassinate North Korean leader Kim Jong-un, and threatened terrorist attacks at cinemas screening the film. After major U.S. cinema chains opted not to screen the film in response to these threats, Sony elected to cancel the film's formal premiere and mainstream release, opting to skip directly to a digital release followed by a limited theatrical release the next day.

United States intelligence officials, after evaluating the software, techniques, and network sources used in the hack, alleged that the attack was sponsored by North Korea. North Korea has denied all responsibility.

5.2.1 Hack and Perpetrators

The exact duration of the hack is yet unknown. A purported member of the Guardians of Peace, who claimed to have performed the hack, stated that they had access for at least a year prior to its discovery in November 2014. The attack was conducted using malware.

Sony was made aware of the hack on Monday, November 24, 2014, as the malware previously installed rendered many Sony employees' computers inoperable, with the warning by GOP, along with a portion of the confidential data taken during the hack. Several Sony-related Twitter accounts were also taken over. This followed a message that several Sony Pictures executives had received via email on the previous

Friday, November 21; the message, coming from a group called "God'sApstls" [sic], demanded "monetary compensation." In the days following this hack, the GOP began leaking yet-unreleased films and started to release portions of the confidential data to attract the attention of social media sites, although they did not specify what they wanted in return.

Other emails released in the hack showed Scott Rudin, a film and theatrical producer, discussing the actress Angelina Jolie. In the emails, Rudin referred to Jolie very negatively because Jolie wanted David Fincher to direct her film *Cleopatra*, which Rudin felt would interfere with Fincher directing a planned film about Steve Jobs.

5.2.2 Threats Surrounding *The Interview*

On December 16, for the first time since the hack, the GOP mentioned the then-upcoming film *The Interview* by name and threatened to take terrorist actions against the film's New York City premiere at Sunshine Cinema on December 18, as well as on its American wide release date, set for December 25. Sony pulled the theatrical release the following day.

> We will clearly show it to you at the very time and places *The Interview* be shown, including the premiere, how bitter fate those who seek fun in terror should be doomed to. Soon all the world will see what an awful movie Sony Pictures Entertainment has made. The world will be full of fear. Remember the 11th of September 2001. We recommend you to keep yourself distant from the places at that time. (If your house is nearby, you'd better leave.)

Seth Rogen and James Franco, the stars of *The Interview*, responded by saying they did not know if it was definitely caused by the film, but later cancelled all media appearances tied to the film outside of the planned New York City premiere on December 16, 2014.

Undoubtedly all of the publicity involved with the North Korean and other threats, and the embarrassing disclosures from the Sony hack, led to a great deal of public interest in the film and the resulting controversy. As a result, many people who would not have been willing to take a risk by going to the theater showing this film decided to purchase and view it in the safety of a streamed video.

Thus, *The Interview* became somewhat of an underground hit, and it is not disputed that many more people saw the film because of the controversy.

The question thus became who might've been the perpetrator of the Sony hack and the resulting actions related to the controversy regarding the release of the film.

5.3 PROFILING MATRICES

With the information given regarding the Sony hack, we can use the information from the series of events to develop a model to try to narrow the consideration of potential perpetrators of this hack as well as their motivations for carrying this out.

Rather than having an approach that will conclusively lead to a definitive answer, the profiling matrix approach will provide the researcher with a means to narrow the potential suspects and corresponding motivations to a point where many suspicions may be effectively eliminated.

First, we will build a list of potential suspects. The beginning step here is to gather as much information as possible regarding candidates to be included or rejected as suspects. From the case study as presented above, most persons would immediately include the North Korean government as a suspect. Along those lines, however, it may be that rather than the North Korean government itself, which may or may not have the requisite technical ability, actors on its behalf might be considered. As we have said, a suspected hacker group that calls itself the Guardians of Peace claimed that it had executed the Sony hack. However, Guardians of Peace had not been previously identified, so it is possible that it was only a pseudonym for some other hacker group. The group Anonymous might also be suspect, since it had claimed, with some reason, that it had perpetrated numerous other hacks in the past.

It is also known through any analysis of world politics that China is the country that has the strongest working relationship with North Korea, and it is also known to have very substantial technical capabilities. So it is not unreasonable to think that perhaps China carried out the attack upon the request of its allied government in North Korea.

But there had also been suspicion related to a technically skilled Sony employee, who is referred to as Lisa in numerous articles. Lisa had been fired by Sony not long before the hack in question, and it was widely known that she had the capability to perform the hack and also may have had various motives for revenge against Sony.

It is also reasonable to consider, as in any competitive environment, that competitors of Sony might have also had the motivation to do damage to Sony's reputation. So one might add to the list of suspects other corporations in the film industry that are Sony's competitors.

At the point where there was considerable discussion as to whether the film *The Interview* might be pulled from being released, or released in a fashion that might diminish its profitability, persons who stood to benefit from the success of the film—for example, the producer and director, or the lead actors—might have a motive in terms of either decreasing or increasing the value of the film. This might in turn result in more or less profitability for the persons directly involved with the film.

A final consideration might be that Sony Pictures itself might have realized that creating considerable controversy over the release of the film, which otherwise might have gone mostly unnoticed, could result in greater profitability for Sony itself.

So this analysis could lead to a list of suspects as follows:

North Korea
Guardians of Peace
Anonymous
China
Lisa
Producer, Director, Actors
Sony Pictures

The next step is to develop a matrix in which the rows and columns will be labeled. To begin, populate each row of this matrix with one of the sets of suspects as described above.

We now have a potential list of suspects; our next task is to try to identify the reasons for the motivations for this series of events. It is conceivable that one could identify a new motivation for which none of the previous suspects (identified by roles) might be identified. Nevertheless, as in many cases, one can usually identify money as a potential motivation. In this case, politics also enters, given the stature of the United States, North Korea, and China.

Any time we identify groups for whom hacking is a primary activity, we should consider if the hack is carried out to demonstrate technical proficiency—or even to try to "advertise" for future business. Revenge is often another motive, certainly in this case given the existence of the disgruntled former employee. And industrial competitiveness is often another potential motivation.

Perhaps one more candidate should be added to this list: given the results that Sony was much more profitable with this film after all of the controversy, it might be considered that stirring the controversy could've been perceived as a potential way to make the series of events beneficial to Sony itself.

Our next step is now to create the columns of the profiling matrix. Thus, we list all of the potential motivations that we have identified, and now we can create a matrix whose rows are labeled by the suspects and columns labeled by the potential motivations. The resulting profiling matrix is displayed in Figure 5.1.

Motivations ->	Politics	Keep the peace	Warfare	Reputation	Start conspiracy	Become famous	Personal vendetta	Money	Disclose Information	adventure	Abuse	Competition
Suspects												
North Korea												
Guardians of Peace												
Iran												
SONY Employees												
WikiLeaks												
Russia												
China												
Microsoft												
Industrial Competitor												
Movie Industry												
Google												
Anonymous												
Lena												
MGM												
Seth Rogen, James Franco												
SONY Pictures												

FIGURE 5.1 Profiling matrix template.

Motivations ->	Politics	Keep the peace	Warfare	Reputation	Start conspiracy	Become famous	Personal vendetta	Money	Disclose Information	adventure	Abuse	Competition
Suspects												
North Korea	43.21	2.50	31.07	18.21	10.00	1.07	10.36	4.29	7.14	0.36	0.36	0.36
Guardians of Peace	12.86	10.36	4.29	41.07	16.43	23.21	9.64	3.21	11.43	3.93	1.07	0.71
Iran	21.26	1.19	12.45	14.12	13.41	2.26	3.76	5.12	6.07	1.07	1.07	0.36
SONY Employees	1.07	11.79	0.89	6.79	3.57	6.43	34.64	20.18	21.25	11.79	2.32	1.07
WikiLeaks	27.90	4.33	4.29	7.65	15.08	1.43	0.00	1.07	41.60	3.02	0.00	1.15
Russia	23.54	0.79	36.70	2.74	30.99	1.94	0.71	5.00	6.62	2.30	0.71	1.87
China	35.78	3.06	22.58	1.02	28.28	3.06	2.14	7.86	3.21	0.00	0.71	0.71
Microsoft	4.12	4.95	1.55	10.55	7.26	0.36	0.00	28.36	5.00	0.00	0.00	15.00
Industrial Competitor	3.57	3.57	0.71	13.57	12.86	1.79	2.86	48.21	2.86	2.14	2.86	15.36
Movie Industry	0.71	7.86	0.00	3.57	6.43	1.43	1.43	15.00	3.57	0.00	0.00	22.86
Google	3.39	9.29	0.89	3.39	5.18	0.00	0.00	16.61	9.64	5.00	0.89	7.14
Anonymous	3.89	1.07	1.07	7.53	17.45	1.38	6.07	4.59	31.54	25.39	2.45	3.65
Lena	0.36	2.14	0.00	17.12	3.81	4.29	30.33	23.14	10.24	0.00	0.00	0.00
MGM	1.43	4.29	0.00	7.14	10.00	0.71	0.00	11.43	0.00	0.00	0.00	14.29
Seth Rogen, James Franco	1.79	13.21	1.79	1.43	2.86	19.29	5.00	20.71	1.79	0.00	0.00	4.29
SONY Pictures	3.93	20.17	2.74	2.14	17.95	5.95	0.00	29.02	2.38	0.00	0.00	8.57

FIGURE 5.2 Profiling matrix for the Sony hack.

Thus, once we have established all of the potential motivations for the attack, we have defined the columns of our profiling matrix:

> The next step in the analysis is to examine each cell in this newly formed matrix, and then, based on all the documentary evidence at hand for the case, estimate the probability that this particular cell, defined by a pair (perpetrator, motivation), might be the most likely guilty party and rationale.

TABLE 5.1
Classification of Malware: The "ABCD" Model for Bad Guys

Level of Criminal	Description	Example
D	Smash-and-grab: no skill, knowledge, resources; sees opportunity and acts immediately	Script kiddies
C	Some limited skills and resources, but little planning in execution	Low-orbit ion cannon
B	Very knowledgeable, some resources, ability to plan	Internet worm
A	Well-organized team, very sophisticated knowledge, lots of resources, only interested in large targets, can plan extensively	Stuxnet

For example, if we consider the pair (North Korea, industrial competition), we would likely assign a near-zero probability. By the same token, the pair (Lisa, politics) would also have a very low probability. This analysis should be undertaken for all combinations of cells in the matrix. In previous courses offered using this profiling matrix technique, students were asked to estimate the most likely probabilities for every cell in the matrix, then to gain a collective judgment; the matrices defined by each student were averaged to give a class response.

What follows is one example developed by one of our classes in the more extensive case of a 11 × 13 or 143-cell matrix (Figure 5.2).

In a later chapter, the method of establishing profiling matrices is extended in a transformation to a two-person, zero-sum game theory model that allows for a solution to be developed from the profiling data.

5.4 "ABCD" ANALYSIS

Another method for analyzing levels of threat in a cybersecurity environment is the so-called "ABCD" approach. This approach, which has its origins in criminal justice theory, attempts to define an attack in terms of its sophistication. As a simplification, potential cyberattacks or the hackers developing or utilizing these attacks are divided into four categories, with the least sophisticated attacks categorized as "D-level," going all the way to the most sophisticated attacks by attackers who fall into the "A-level" category (Table 5.1).

We will explore this classification scheme in greater detail in Chapter 25.

REFERENCES

Alvarez, E. 2014. Sony Pictures hack: The whole story. *Engadget*, https://www.engadget.com/2014/12/10/sony-pictures-hack-the-whole-story/

Soffer, K. 2016. The big question about why police pull over so many black drivers. *Washington Post*, July 8. https://www.washingtonpost.com/news/wonk/wp/2016/07/08/the-big-question-about-why-police-pull-over-so-many-black-drivers/?utm_term=.7346a524986f

Sony Pictures Digital Productions Inc. 2018. http://www.sonypictures.com/
Turing, A. M. 1950. Computing machinery and intelligence. *Mind: A Quarterly Review of Psychology and Philosophy*, LIX(236), 433–460.

PROBLEMS

5.1 Find an example beyond "driving while black" with the generic assumption of identifying with an ethnic or racial stereotype.
5.2 Identify some values and some dangers of profiling.
5.3 Can you find a tragic outcome of profiling?
5.4 Assess the reviews of movie *The Interview* related to the Sony hack.
5.5 Have there been any prosecutions for the Sony hack? If so, describe them. If not, why not?
5.6 Has "Lisa" ever been identified?
5.7 Identify your own candidates as suspects and motivations for the Sony profiling case.
5.8 Construct profiling matrices for:
   a. The "Nine Iranian Hackers" case. See: https://www.justice.gov/opa/pr/nine-iranians-charged-conducting-massive-cyber-theft-campaign-behalf-islamic-revolutionary
   b. The WannaCry Ransomware case. See Chapter 4.8.
   c. Stuxnet. See Chapter 4.6.
5.9 Construct an ABCD description for the classification of potential bank robbers.

6 Hack Lab 1 *Social Engineering Practice: Who Am I?*

Throughout this book you will discover a number—in particular, four—of what we call Hack Labs. These labs are designed to give students practical experience in dealing with a number of cybersecurity issues that are of critical concern in the protection of computing environments.

The other purpose for these labs is that it is not necessary, but there could be a supportive physical computer lab to carry out these projects. They can also be done on a student's own computing equipment and do not have to be done within a fixed lab period.

When these have been offered by the authors, they have usually allowed the students a week to carry out the research and submit the results.

6.1 HACK LAB 1: SOCIAL ENGINEERING: FIND COOKIE'S PASSWORD

The first of these labs deals with an approach used by many hackers that is sometimes called social engineering or dumpster diving. Most naïve or even not-so-naïve computer users do not realize how an attacker may try to gather sufficient information about a target in order to be able to make an educated guess about the target's password.

It should also be remembered by those who are responsible for maintaining security on a multiuser system, such as a university's primary network, that an attacker does not need to capture thousands of passwords in such a system: only one will probably suffice to meet the hacker's need.

This lab will ask students to try to determine Cookie Lavagetto's password. It is given that Cookie, as a somewhat careless user, has a password that is constructed from two separate pieces of information about Cookie and his family. In the data sheet that follows, note that there are 21 distinct data points regarding Cookie and his family: family names, addresses, children, their ages and birthdays, their pets, and pet names. This is information that most dumpster divers should be able to accumulate.

Incidentally, the inspiration for this example is from the best film ever about computer hacking, the 1984 thriller called *War Games* where the protagonist is a teenager who is trying to find a password to break into a gaming computer. According to the plot, it is actually a defense department computer that could start World War III invented by a computer scientist named Falken. How the teenager discovers Falken's secret password is not only a fascinating part of the film, but also a marvelous lesson in the techniques of password finding or dumpster diving.

Rather than timing this lab, it has proven to be more instructive to set a limit on the number of tries to find the password for each student. The detail of carrying this out can be one of two approaches: (1) the instructor can create a password mechanism with a database of all the possible combinations, and then accept or reject the student's try. The instructor can also record how many tries the student has made up to a preset limit, or (2) the instructor can find a website that allows the user to create an account and set a password that might be created from two of Cookie's data points, for example: oreo123 or woofieebbetts. It would be reasonable to accept the principle that the passwords are non–case sensitive.

With 21 data points and a rule established that a password can be no more than a combination of two of the given data points, there are $(21 \times 20)/2 = 210$ possibilities. You might limit the lab to trying, say, 30 combinations.

6.2 COOKIE'S DATASET: INSTRUCTIONS TO STUDENT

Cookie Lavagetto

Spouse: Cookette Lavagetto

Residence: 200 Ebbetts Field Road, Brooklyn, NY 10093

Children: Oreo Lavagetto, age 14; Gingersnap Lavagetto, age 12; Tollhouse Lavagetto, age 7

Pets: Spot, Woofie, George

Birthdays: Cookie 1/23, Cookette 8/17, Oreo 3/9, Gingersnap 11/28, Tollhouse 4/3

Cookie's email: The instructor creates this.

How to play: Log in to the given account.

Try to guess a password. Suggestion: Before entering electronically, construct and write down 10 guesses. Then see if you've guessed the correct password.

The correct password is some combination of the data above. It will be no more than 14 characters.

If you are successful in guessing the password, you have hacked Cookie's account. Submit as the assignment (by email) the successful password you have entered. *Like any good hacker, keep your successful password to yourself.*

The Hack Lab will run for 1 week or until you make 30 attempts. I will change the password three times during the week, so you may have a total of four correct solutions. You will get full marks if you get any two of the four.

PROBLEMS

6.1 Cookie has 21 data points. If a password is constructed from a combination of two data points, how many tries will it take to guess the correct password?

6.2 Construct a separate example. Allow a classmate five guesses. How many of your classmates will find the password?

6.3 Can you find a real-world example of dumpster diving?

6.4 How did the high school student in the movie *War Games* discover Falken's password? What was this password and why would Falken have chosen it?

7 Access Control

7.1 ACCESS CONTROL

Our first step in protecting a computing environment or cyberenvironment is to establish methodologies for determining how access may be gained to our environment. We usually divide this concern into two components that we call *authentication* and *authorization.*

Our concern in providing authentication is basically to answer the question "Who are you?" In other words, this means the establishment of a mechanism for determining whether a party wishing to gain access is allowed to enter the system. In this case, the party in question might be either a human or a machine. The authentication process is initiated by that external party, and our system must respond appropriately.

The second aspect of access control is called authorization. In other words, once an external party has been authenticated, questions may arise as to whether that party has the authority to perform certain tasks in this new environment. In other words, the authorization question might be "Are you allowed to do that?" And so, our system must have a methodology for enforcing limits on actions.

7.2 AUTHENTICATION

The process of authentication begins with a request from the external party, followed by a challenge from our system, which usually can be divided into one of three approaches: the challenge can be based on what we usually describe as

Something you know
Something you have
Something you are

7.3 SOMETHING YOU KNOW: PASSWORDS

The something you know is usually thought of as a password. Something you have may be some physical device such as a key, smart card, or some other token. And something you are is usually described as a biometric, in other words, your fingerprints, the image of your face, the scan of your retina, your method of walking or gait, or your DNA, among others.

Consider first the example of a UNIX or LINUX system (Kernighan, 1984). The system might display:

Login:

As simple as this may seem, the choice of the password is one of the greatest risks in providing good cybersecurity.

Thus, there is a long history of users creating good or bad passwords, and this has been one of the biggest problems in the world of cybersecurity. The concept is that the user chooses a good password—in other words, one that is hard for an outsider to guess—and that will foil an outsider from making a successful guess.

The first step in creating—or analyzing—a password system is to know the rule for allowable passwords. Every password system has allowable types of symbols or characters to be typed. Examples are digits { 0, 1, 2, ..., 9 }, letters of the alphabet (lowercase) { a, b, c, ..., z }, special (typable) symbols { #, $, %, ⊥, &, ?, ! }, or combinations of these. The set of these that is being used we will designate as c, for the character set.

The second part is how many of these symbols may or must be typed in for a legitimate password. For the moment, consider that this must be a fixed number, n.

It is important to know, therefore, how many possible passwords there can be. Since there are c choices for each entry, and there must be n of them, the total number of potential passwords is c^n.

Example: For many ATM PINs, four digits are required, and there are 10 potential digits. So, c = 10, n = 4, and the total number of possible PINs is $c^n = 10^4 = 10{,}000$.

Example: In many older password systems, seven characters needed to be entered, each one a lowercase letter. Thus, $c^n = 26^7 = 8{,}031{,}810{,}176 = 8.03 \times 10^9$ or just over 8 billion possible passwords.

Let's call this value the "password set," $p = c^n$. This value also indicates the challenge to someone trying to obtain someone else's password.

Since the system itself will usually instruct a potential user as to the password rules, the hacker trying to determine a user's password will know c and n and thus can calculate p. So, the most obvious hacker's approach is usually called "brute force" or "exhaustive search"—try all the potential passwords.

Imagine that all the potential passwords are listed one after the other and the hacker simply goes down the list. The hacker has an equal possibility of hitting on the first try as not hitting until the very last. Thus, by a simple probability argument, on average, it will take going halfway down the list to find a hit. Therefore, the work effort necessary for the hacker is essentially t = p/2 tries in order to find the password. In our first example above, this means t = 10,000/2 = 5,000 tries; in the second case, t = 8,031,810,176/2 = 4,015,905,088 tries.

Of course, the hacker is not limited to a brute-force strategy. We'll return to this analysis later.

7.4 TOKENS: WHAT YOU HAVE

Devices that you physically possess are becoming less and less common in contemporary computing environments. The reason is simply that if you have some form of key, token, smartcard, or other physical device, and if it is simply lost or stolen and only that physical device is necessary for entry or authorization in some environment, then essentially the door is left wide open.

Consequently, more and more the physical device is used as part of what is called "two-factor authorization." In other words, the physical device itself that you have is not sufficient to allow entry or provide authorization in a given environment.

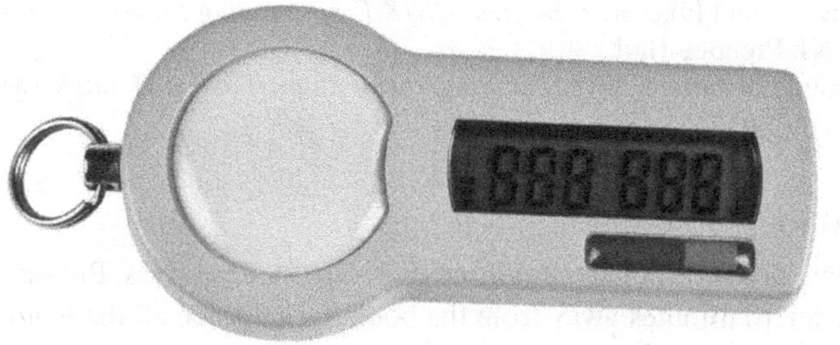

FIGURE 7.1 RSA SecurID token.

Normally, the physical device is combined with a second factor, normally a password. One very widespread example is the ATM card, where the card itself is a physical device, but in order to use it at an ATM, you need a password, which we typically call a PIN—usually a four-digit number.

A more clever device is the so-called RSASecurID security token (CDW, 2018), which is synchronized in a multiuser system with the time that the token is activated. This device has a six-digit display, which changes every 60 seconds. The user with the token must enter the displayed six digits concurrently with the entry of a password. The central system uses an algorithm that can determine to the minute what the token's six-digit readout should display, and the user must enter the six digits to match what the system calculates (Figure 7.1).

7.5 BIOMETRICS: WHAT YOU ARE

The field of biometrics has been in existence much longer than the computer era. Perhaps one of the most common biometric measurements—what you are—is the fingerprint (NIJ, 2018). The study of classification of humans by fingerprint dates back to the nineteenth century. Other more contemporary biometric measurements include facial recognition, hand recognition, retinal patterns, and DNA.

Although many of these biometric measures are sufficiently accurate to uniquely determine any individual in the entire world's population, there are number of problems with biometric measures. For example, passwords can be easily changed by a user in conjunction with system management. Such is not the case with most biometric measures. Furthermore, some potential users may find that the requirement to provide biometric measures is an invasion of privacy. For one example, detecting iris or retinal patterns involves shining a laser into the subject's eye, a process many people may feel is not only invasive, but physically harmful (it isn't!).

REFERENCES

CDW Inc. 2018. RSA SecureID. https://www.cdw.com/content/dam/CDW/brands/rsa/securid-access-guide.pdf?cm_ven=acquirgy&cm_cat=bing&cm_pla=S3+RSA&cm_ite=RSA+SecurID+E&s_kwcid=AL!4223!10!73873530364386!73873502272841&ef_id=VqMBwgAABUqYPytc:20181106160059:s

Kernighan, B. W. and Pike, R. 1984. *The UNIX Programming Environment.* Upper Saddle River, NJ: Prentice-Hall.
National Institute of Justice. 2018. *The Fingerprint Sourcebook.* NIJ. https://www.ncjrs.gov/pdffiles1/nij/225320.pdf

PROBLEMS

7.1 Here are several lists. Examine each one for 2 minutes. Put the list away. After 10 minutes away from the book, write down all the words you can remember without looking back.

6-Letter Words	6-Character Symbols	8-Letter Words	10-Letter Words
whirrs	kjmarj	calmness	countermen
recces	zlxgqc	fruition	boundaries
bourgs	uzsfsh	reawaken	inimitable
cowboy	gmgbxx	pretense	democratic
creped	nsucha	incurred	repackages
coffee	tylkxz	relocate	cellophane
alters	cyauml	pinching	watchwords
guests	aayljd	severity	supermodel
choker	sinmaj	throated	fellowship
mazing	dfasfy	drilling	yourselves

7.2 Check your score. Repeat the problem for each list. Compare your scores for each list.

7.3 Which of the following pairs of password schemes provides the greater security in terms of the largest number of legal passwords in the scheme?
   a. U: (c,n) = (26,8) or V: (c,n) = (20,9)
   b. W: (c,n) = (52,8) or X: (c,n) = (72,7)
   c. Y: (c,n) = (62,13) or Z: (c,n) = (20,18)

7.4 Suppose the password set size is 10^9 = 1 billion. Find a minimal value for c when n = 7. Find a minimal value for n when c = 52.

7.5 Suppose we have the following password authentication schemes. Indicate the solutions for each case in the matrix below, assuming a brute-force attack.

Case	Scheme	Total Number of Attempts Necessary to Find a Password	Expected Number of Attempts Necessary to Find a Password
A	Exactly 6 alphanumeric characters, case sensitive		
B	Any word in the Oxford English Dictionary		
C	Fifteen characters, which all must be vowels (i.e., a, e, i, o, u, y)		
D	A 4-digit PIN (numeric characters only)		

(Continued)

Case	Scheme	Total Number of Attempts Necessary to Find a Password	Expected Number of Attempts Necessary to Find a Password
E	Six characters, which must have at least one in each category: non–case sensitive letters, numbers, and one of the following special symbols { ! # $ % ⊥ & * ~ - / }. ("{ }" represents a set notation, not two of the special characters)		
F	Exactly 10 alphabetic characters, non–case sensitive		

7.6 In the film "War Games," one scene shows a massive computer, the WOR, that breaks a password as numbers shown as tumblers revolve at a high speed and the computer locks first on the first character, then on the second character, and so on. Discuss this method to break a password.

7.7 Consider the RSA token. With a six-digit display, changing every minute, how long will it take until the display repeats?

7.8 Describe a model for "three-factor authentication."

7.9 Suppose you are a merchant and decide to use a biometric fingerprint device to authenticate people who make credit card purchases at your store. You can choose between two different systems: System A has a fraud rate (false ID accept as correct) of 1% and an insult rate (bogus ID accepted as real) of 5%, whereas System B has a fraud rate of 5% and an insult rate of 1%. Which system would you choose and why?

7.10 We have a biometric registration system enrolling four subjects, Alice, Bob, Caroline, and Dan. The database reads:

Alice	1000	1110	1010
Bob	0111	0100	1101
Caroline	0101	0101	0101
Dan	1110	0010	0110

A person arrives for recognition and the value x is read:

x	0101	0101	1101

If we know this is a legitimate subject, who is it and why?

8 The First Step
Authorization

Now we assume that an external user has satisfied the requirements for authentication and has entered a system. Although for many of us, we may have a personal computer with access only for one person, more and more the norm in a cyberenvironment is that there may be multiple parties present in the same environment with distinct sets of resources (such as files or permission to run various types of software).

Furthermore, the environment with numerous parties participating at the same time might be a single, multiuser computer system (that in other times we might've called a mainframe) or conceivably a networked environment of many small or single-user computers communicating and sharing resources managed by the network software.

In either case, we may assume that there is a computing environment capable of maintaining many users who have ownership of or access to many types of resources.

In this context, the problem of authorization must be addressed. Suppose a user requests access to a specific file. What is the mechanism to ensure that that particular user has the authorization to either read, write, or delete that file, no matter where in this environment this file or resource resides? Over time, there have been many approaches to this problem. Two standard methods are known as the maintenance of access control lists or capabilities.

8.1 LAMPSON ACCESS CONTROL MATRIX

An early approach that at least describes the problem was a methodology known as Lampson's access control matrix (Lampson, 1971). In such a system, the entities in the multiuser environment are called subjects and objects. Subjects are the entities that will request a specific resource, while the objects are the resources to be requested. Examples of subjects are users, executable programs that require data, or other types of processes. Examples of objects are files, disk drives, executable programs to be run, or printers.

It is not unheard of in a complex system that there may be in fact thousands of subjects or objects. In the Lampson model, a matrix is maintained where each subject has a row in the matrix, and each object has a column. When an action occurs in the system, it is likely that there will be a request by one of the subjects for a resource that constitutes one of the objects. The management system or operating system must determine whether this request is legitimate. The system will determine this by examining the element in the matrix corresponding to that subject and the corresponding object, and the content of that matrix element will determine whether the request is to be honored or denied (Table 8.1).

Users with a familiarity with the UNIX environment may recall that its approach to such management is to attach to each object a string that contains something that

TABLE 8.1
The Lampson Matrix

		Objects	(Resources)	Index the Column's		
		Operating Systems	Inventory Program	Inventory Data	Equipment Data	Payroll Data
Subjects	Mary	rx	rx	r	---	---
(users)	George	rx	rx	r	rw	rw
Index the rows	Susan	rwx	rwx	r	rw	rw
	Inventory program	rx	rx	rw	rw	rw

looks like "-rwxrwxrwx." The meaning of this string is that the object or file may be read (r), written (w), or executed (x). If the first position is "-" as in the example, the object is a file; the other possible value in the first position is "d," denoting that the item is a directory. In the example, the repetition of the "rwx" in the second to tenth symbols indicates the permissions applied to, respectively, the user him- or herself, a group to which that user is attached, and the "world" for all users with access to the system. As a further example, a file with the permissions "rwxrw-r--" may be read, written, or executed by the user; read or written by the user's group; and only read by all legitimate users in the system.

The Lampson model, as indicated above, is often in these times very hard to manage, as with even 1000 subjects and 1000 objects, the matrix will have one million entries to search to determine the legitimacy of any transaction.

8.2 SECURITY LEVELS

Complex modern environments tend to have multiple ways of representing the level of access going beyond the UNIX model described above. Many of us will be familiar with the specific levels of access in a government or military system. In perhaps its simplest form, these levels can be described as Top Secret, Secret, Confidential, and Unclassified.

In such a system, these levels are applied to each subject and object. Then the manner of accepting or rejecting a specific request will follow this hierarchy:

Top Secret > Secret > Confidential > Unclassified

Systems implementing such a hierarchy are normally called multilevel security models (MLSs). An early method of implementing such a model was the so-called Bell-LaPadula model, which seems contradictory at first, but in fact has served as the basis in many modern computing environments (Bell and LaPadula, 1973). The simplest form of the Bell-LaPadula model follows this set of rules:

- Classifications apply to objects
- Clearances apply to subjects

- U.S. Department of Defense uses four levels of classifications/clearances:
  - Top secret
  - Secret
  - Confidential
  - Unclassified

The Bell-LaPadula (BLP) security model is designed to express the essential requirements for an MLS. The primary concern of BLP deals with confidentiality, in order to prevent unauthorized reading. Recall that O is an object, and S a subject.

Object O has a classification, whereas subject S has a clearance. The security level is denoted L(O) for objects and L(S) for subjects.

The BLP rules consist of the

Simple security condition: S can read O if and only if $L(O) \leq L(S)$
**-Property (star property)*: S can write O if and only if $L(S) \leq L(O)$

As a shorthand term, this is often referred to as "no read up, no write down."

8.3 PARTIAL AND TOTAL ORDER

It is normal in many authorization systems for the user to have not only a security level, as discussed above, but also a secondary level of authorization involving a subset of the users at a given security level. For example, a number of users with Secret clearance may be assigned to work together on a single project, but all other users with Secret clearance not involved in this project have no need for the information that is contained in the development of the project. This is the so-called "need-to-know" principle. Thus, in this case, a user working on the project will by necessity have Secret clearance, but also need to be a member of the group assigned to the specific project. Therefore, the complete security status for anyone in this environment will require both the level of clearance and also the listing of the groups to which the user belongs.

Thus, when a request is made by a subject to access a given object, the system must check both respective security levels. The overall security level is described as a "total order." In other words, it is always possible to determine the access: a Confidential user may never read a Secret object; a Top Secret subject may read an Unclassified file; a Secret user may write to a Top Secret object.

However, the permissions involved in the group structure form what is known as a "partial order." The easiest way to think of a partial order is in terms of a Venn diagram. Suppose we have sets in a Venn diagram as displayed in Figure 8.1. One set in the diagram is contained in another if it is completely enclosed; in other words, one set is a subset of the other. We usually use the symbol $\subseteq$ to indicate this set inclusion (Wolfram, 2018).

If, however, two subsets overlap but neither is completely enclosed in the other, then no set inclusion exists. For this reason, we describe this ordering as a partial order.

Consider this example of a project where the teams are divided into groups called { Red, Blue, Green, Yellow, Brown }, user A belongs to groups { Red, Green, Brown }, and object B belongs to groups { Blue, Green, Yellow }. Then neither a request for A

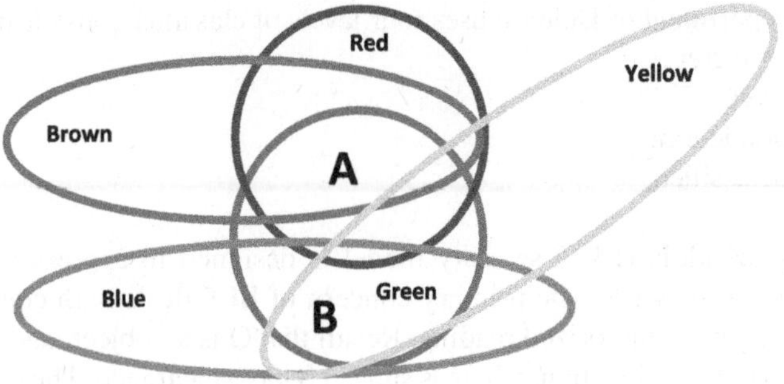

FIGURE 8.1 Venn diagram for { Red, Blue, Green, Yellow, Brown }.

to read B nor a request for A to write to B will be honored, because the subset for A is not contained in the subset for B, nor vice versa (Figure 8.1).

8.4 COVERT CHANNEL

Despite the protections involved in the access control system, there can always be the existence of a mechanism to "get around" these protections. These are typically what are called "covert channels." Covert channels are often defined as communication paths that are not intended as such by system designers (Techopedia, 2018).

Here is one example that could exist in a multilevel security system. Suppose that a user, George, has Top Secret clearance, and another user, Phyllis, has Confidential clearance. Furthermore, suppose that the space for all files is shared. In this case, George creates a file called ABC.txt, which will be assigned Top Secret level since it was created by George with that clearance. Of course, Phyllis cannot read that file, having only Confidential clearance. However, in order for George to communicate a covert message bit by bit to Phyllis, Phyllis checks George's directory every minute. If in the first minute, Phyllis sees the existence of ABC.txt, she records a "1" bit. If, one minute later, George deletes the file and Phyllis checks again—by not seeing the name of the file in the directory—she will record a "0." Then, George may create the file again, so in the next minute, Phyllis will see the existence of the file and record a "1." Thus, over a period of, say, 60 minutes, George can "leak" essentially a message of 60 bits.

In a similar fashion, the print queue, ACK bits, pings, or unused locations in a TCP network packet could be used for similar covert messages (Figure 8.2).

A real-world example was described to us at the Robben Island prison in South Africa where Nelson Mandela was held as a political prisoner for over 20 years. The prisoners were separated into small groups, but for their recreation they were allowed to play tennis with broken-down tennis balls. The prison guards thought that the prisoners were lousy tennis players because they kept hitting the balls into the next cell area. But the prisoners would hide messages into the broken tennis balls, essentially serving as a covert channel for the messages inserted into the tennis balls.

Consider one further example of a covert channel. Assume that we have a 100-MB Top Secret file, stored in its unencrypted fashion at the Top Secret level.

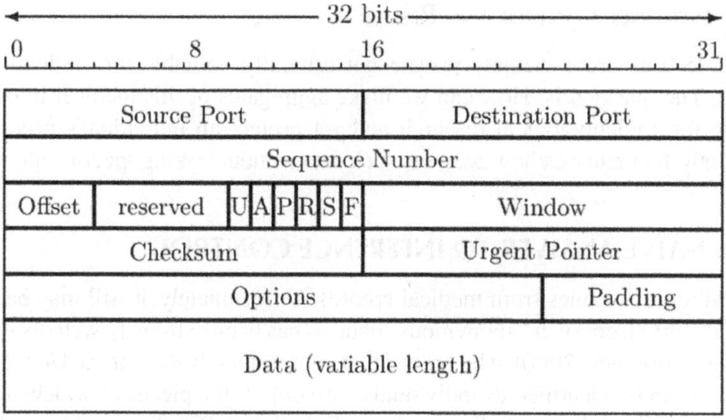

FIGURE 8.2 Example of a TCP packet.

However, the encrypted version may be stored at the Unclassified level for the purposes of transmitting it from one system to another. This presents no security risk because we would assume that the encrypted version is useless to someone discovering this version without the key. However, the key, which would of course be stored at the Top Secret level, may only occupy perhaps a few hundred bits. Thus, a method such as that described above, leaking one bit at a time from the Top Secret level to the Unclassified level, could make the complete key available at the Unclassified level in just a few minutes, and pass on the ability to decrypt the 100-MB file.

8.5 INFERENCE CONTROL

A major application for many is use and access in a major database, such as a corporate database, or perhaps a university database. The purpose of a database is to allow for multiple users to read or modify information from the database according to access rules such as those described above. However, access to a database is not frozen at a moment in time. Any user with the ability to submit a database query may be able to inadvertently (or perhaps intentionally) gain information that the user may be prohibited from accessing directly. Consider the following example from a university database:

Question: What is average salary of female psychology professors at XYZ University?
Answer: $95,000

Question: How many female psychology professors are there at XYZ University?
Answer: 1

As a result of these two queries, specific information has leaked from responses to general questions!

8.5.1 Inference Control and Research

Medical records are extremely private but critically valuable for epidemiological research. The question is: How can we make aggregates of this medical information available for epidemiological research and yet protect an individual's privacy? Or, equivalently, how can we allow access to such data without leaking specific information?

8.6 A NAÏVE ANSWER TO INFERENCE CONTROL

We could remove names from medical records. Unfortunately, it still may be easy to get specific info from such "anonymous" data, as has been extremely well documented by LaTanya Sweeney (2000), who proved that approximately 87% of the United States' population can be identified as individuals with only three pieces of widely available data, { gender, birthdate including year, and five-digit ZIP code of residence }.

Therefore, removing names is not enough.

8.7 SLIGHTLY LESS NAÏVE INFERENCE CONTROL: QUERY SET SIZE CONTROL

In this approach to responding to a query, don't return an answer if the set size of the responses is too small.

A well-established principle is the "N-respondent, k% dominance rule." In other words, do not release the statistic if k% or more responses are contributed by N or fewer records. This will fail in some cases, for example:

Example: The average salary in Bill Gates' neighborhood

However, this process can be defeated by two cooperating individuals collaborating on submitting individual queries designed to get around this rule.

8.7.1 Randomization

You can add a small amount of random noise to data. However, this works with fields like weight, age (such as 45.2 years), or height, but could not be used with ZIP code, gender (represented by a 0 or 1), or ethnicity—in other words, integer values and not rational or real numbers.

8.8 FIREWALLS

A firewall must determine what to let into the internal network and/or what to let out. In other words, the firewall is designed to provide access control for a network (Cisco, 2018).

You might think of a firewall as a gatekeeper. To meet with an executive, first contact the gatekeeper. The gatekeeper decides if the meeting is reasonable; also, the gatekeeper filters out many requests.

Suppose you want to meet the chair of the psychology department. The gatekeeper does some filtering. But if you want to meet the president of the United States? The gatekeeper does lots of filtering.

There are several categories of firewalls:

- Packet filter
- Stateful packet filter
- Application proxy
- Personal firewall—for single user, home network, etc.

8.8.1 The Packet Filter

This firewall operates at network layer, and it can filter based on

- Source IP address
- Destination IP address
- Source port
- Destination port
- Flag bits (SYN, ACK, etc.)
- Egress or ingress

8.9 INTRUSION DETECTION

In spite of firewalls or other methods of intrusion prevention, bad guys will often get into a system. Thus, if we make the reasonable assumption that our system will be penetrated, we should look at systems that will determine if such a penetration has taken place. Such systems are generally known as intrusion detection systems (IDSs) (Cyberpedia, 2018).

IDSs are designed to detect attacks by looking for "unusual activity." IDS is currently a very *hot* research topic.

Who is a likely intruder into our environment? It may be an outsider who got through a firewall, or it may be an evil insider. What do intruders do? They may launch well-known attacks, launch variations on well-known attacks, launch new or little-known attacks, or use a system to attack other systems.

There are two primary intrusion detection approaches. They are *signature-based IDSs*, and *anomaly-based IDSs*.

8.10 SIGNATURE INTRUSION DETECTION SYSTEM

Here is one example of the signature IDS:

Failed login attempts may indicate that a password-cracking attack is underway. The signature IDS could use the rule "N failed login attempts in M seconds" as a *signature*.

If N or more failed login attempts occur in M seconds, the signature IDS warns of a potential attack. Note that the warning is specific. The administrator knows that an attack is suspected, and can therefore verify that it is an attack (or a false alarm).

Using this assumed attack, the IDS warns whenever N or more failed logins occur in M seconds. However, the system must set N and M so that false alarms are not common. We can do this based on normal behavior, but if an attacker knows the signature, he or she can try N – 1 logins every M seconds! In this case, signature detection slows the attacker, but might not stop him or her.

Disadvantages of signature detection include:

Signature files must be kept up to date
Number of signatures may become large
Can only detect known attacks
Variation on known attack may not be detected

8.11 ANOMALY DETECTION SYSTEM

Anomaly detection systems look for unusual or abnormal behavior. In trying to develop such a system, there are (at least) two challenges: First, what is "normal" for this system? Then, how "far" from normal is abnormal? This approach implies that we must use some statistical measurement, for example, standard deviation and variance. In general, the *mean* defines normal, and the standard deviation or *variance* indicates how far abnormal lives from normal.

The question then becomes: How do we measure "normal"? We need to analyze the system during "representative" behavior, not during an attack, or else the attack will seem normal.

How do we measure "abnormal"? Of course, abnormal is relative to some "normal," and abnormal indicates a possible attack.

Suppose we monitor the use of three commands:

open, read, close

Under normal use, we observe that Ellen performs the following operations in this sequence frequently:

open, read, close, open, open, read, close, …

Of the six possible ordered pairs, four pairs are "normal" for Ellen:

(open,read), (read,close), (close,open), (open,open)

Can we use these sequences to identify unusual activity?
We monitor use of the three commands
open, read, close
If the ratio of abnormal to normal pairs is "too high," we warn of a possible attack.
We could improve this approach by

Also using the expected frequency of each pair
Using more than two consecutive commands

TABLE 8.2
Anomaly Detection for Four Types of Usage at Time t_1

H_0	H_1	H_2	H_3
0.10	0.40	0.40	0.10

TABLE 8.3
Anomaly Detection for Four Types of Usage at Time t_2

A_0	A_1	A_2	A_3
0.10	0.40	0.30	0.20

Including more commands/behavior in the model
Using more sophisticated statistical discrimination

Consider now that we monitor Ellen based on the frequencies of her accessing four particular files $\{F_1, F_2, F_3, F_4\}$. Over time, suppose Ellen has accessed file F_n at rate H_n (Table 8.2).

Recently, Ellen has accessed file F_n at rate A_n (Table 8.3).

Is this "normal" use? To estimate this, we compute the distance from the mean:

$$S = (H_0 - A_0)^2 + (H_1 - A_1)^2 + \cdots + (H_3 - A_3)^2 = 0.02$$

And we make a judgment call that if $S < 0.1$, the behavior is normal. In this case, $S = 0.02 < 0.1$, so this is normal.

Problem: How do we account for use that varies over time?

To allow "normal" to adapt to new use, we update long-term averages as

$$H_n = 0.2A_n + 0.8H_n$$

Then H_0 and H_1 are unchanged, $H_2 = 0.2 \times 0.3 + 0.8 \times 0.4 = 0.38$, and $H_3 = 0.2 \times 0.2 + 0.8 \times 0.1 = 0.12$

And the long term averages are updated as Table 8.4.

The updated long-term average is Table 8.5.

And the new observed rates are Table 8.6.

TABLE 8.4
Anomaly Detection Updating Averages

H_0	H_1	H_2	H_3
0.10	0.40	0.38	0.12

TABLE 8.5
Anomaly Detection Long-Term Average

H_0	H_1	H_2	H_3
0.10	0.40	0.38	0.12

TABLE 8.6
Anomaly Detection New Observed Rates

A_0	A_1	A_2	A_3
0.10	0.30	0.30	0.30

Is this normal use?
Compute $S = (H_0 - A_0)^2 + \cdots + (H_3 - A_3)^2 = 0.0488$
Since $S = 0.0488 < 0.1$, we consider this normal
And we again update the long-term averages by $H_n = 0.2A_n + 0.8H_n$

REFERENCES

Bell, D. E. and LaPadula, L. J. 1973. *Secure Computer Systems: Mathematical Foundations.* MITRE Technical Report 2547, Volume I, March 1.

Cisco Systems, Inc. 2018. https://www.cisco.com/c/en/us/products/security/firewalls/what-is-a-firewall.html

Cyberpedia. 2018. Palo Alto Networks. https://www.paloaltonetworks.com/cyberpedia/what-is-an-intrusion-detection-system-ids

Lampson, B. W. 1971. *Protection.* Proc. Fifth Princeton Symposium on Information Sciences and Systems, Princeton University, March, pp. 437–443, reprinted in Operating Systems Review, 8(1) January 1974, pp. 18–24.

Sweeney, L. 2000. *Simple Demographics Often Identify People Uniquely.* Pittsburgh: Carnegie Mellon University, Data Privacy Working Paper 3.

Techopedia. 2018. https://www.techopedia.com/definition/10255/covert-channel

Wolfram MathWorld 2018. http://mathworld.wolfram.com/VennDiagram.html

PROBLEMS

8.1 A system contains the following:

Subjects (with classifications and compartments):

Subject	Classification	Compartments
John	Top Secret	{ Accounting, Operations, Development }
Mary	Secret	{ Accounting, Operations, Production, Marketing }
David	Confidential	{ Production, Development }
Ann	Unclassified	{ Accounting, Operations, Marketing, Development }

Objects (with clearances and compartments):

Object	Clearance	Compartments
Payroll	Top Secret	{ Accounting, Operations, Production }
Inventory	Secret	{ Operations, Production, Marketing }
Shipping	Confidential	{ Accounting, Operations, Production, Development }
Media	Unclassified	{ Accounting, Operations, Production, Marketing }

With a Bell-LaPadula mandatory access control system, indicate whether each of these requests will be accepted or rejected:

Subject	Command	Object	Accept or Reject
John	Read	Inventory	
Ann	Write	Payroll	
Mary	Read	Media	
Ann	Read	Shipping	
David	Write	Shipping	
John	Write	Media	
David	Write	Shipping	
Mary	Read	Inventory	

8.2 The Pretty Good Software Company has an unusual MLS security system. It has classifications/clearances in this hierarchy: Extremely Top Secret, Top Secret, Pretty Secret, Secret, Very Confidential, Confidential, Somewhat Confidential, Public. Also, there are the following compartments: Software, Graphics, DataBase, Operating_Systems, Artificial_Intelligence, and Security.

a. What are the total number of possibilities for assignments of classifications/clearances and compartments to a subject or object?
b. If the company has implemented a Bell-LaPadula policy, what will be the result of the following attempted accesses?

A subject with Pretty Secret and {Software, Graphics, DataBase, Artificial_Intelligence } tries to read an object that is Very Confidential and { DataBase, Software, Artificial_Intelligence, Graphics }.

A subject with Somewhat Confidential and { Graphics, DataBase } tries to write to an object that is Top Secret and { Software, DataBase, Operating_Systems, Artificial_Intelligence, Security }.

8.3 The only compartments defined for the system described in Question 1 are Accounting, Operations, Production, Development, and Marketing. How many possible allocations of compartments to subjects or objects can exist with *exactly* three compartments?

8.4 Use the same system as above. Suppose some subject with a three-item compartments list tries to read an object with a *different* three-item compartments list; the subject's classification is Secret, and the object's clearance is Confidential. Will this be accepted or rejected?

8.5 We have a universe U of objects that are the letters of the Roman or English alphabet, U = { a, b, c, …, z }. In this universe, we have objects (words) that define subsets whose elements are the letters that make up the word. For example, the subset for the word elephant is Elephant = { a, e, h, l, n, p, t }, and Football = { a, b, f, l, o, t }. (Ignore capitalization and repetition of letters.) Construct the Venn diagram that displays the sets corresponding to the words Onomatopoeia, Penicillin, Comatose, Lugubrious, and Syzygy.

8.6 The U.S. Department of Defense, in understanding that covert channels can never be completely eliminated, established a DoD guideline to allow covert channels whose capacity is no more than 1 bit/second. Suppose a malicious leaker (Trudy) adheres to these guidelines. What is the shortest period of time that Trudy can take to leak a 10-Kb file?

8.7 Inference control: You can find a good deal of U.S. city population data at: https://factfinder.census.gov/faces/tableservices/jsf/pages/productview.xhtml?src=bkmk. Formulate a query that will give you all cities with a population less than 100,000 at the last census but greater than 100,000 at the current time.

8.8 a. Suppose that at the time interval following the results in the chapter and Table 8.1 (also below), Ellen's file use statistics are $A_0 = 0.21$, $A_1 = 0.15$, $A_2 = 0.28$, and $A_3 = 0.36$. Is this normal for Ellen? Give the updated values for H_0 through H_3.

b. What if Alice's file use statistics are $A_0 = 0.14$, $A_1 = 0.32$, $A_2 = 0.26$, and $A_3 = 0.28$?

8.9 In the time interval following the results in the textbook and Table 8.4 (also below), what is the largest value that Trudy can have for A_3 without triggering a warning from the IDS? Also, give values for A_0, A_1, and A_2 that are compatible with your specified A_3 value and compute the statistic S, using the H_i values in Table 8.6.

Table 8.4

H_0	H_1	H_2	H_3
0.10	0.40	0.38	0.12

8.10 Assume that we have an intrusion detection system that monitors four activities for a user that we label H_0, H_1, …, H_3. Suppose also that the IDS system recognizes an anomaly if the statistic S calculated from the difference of squares exceeds 0.2. Given the stored table of H values and the current activity table of A values below, does the current activity represent an anomaly? Why or why not?

H_0	H_1	H_2	H_3
0.37	0.38	0.13	0.12

A_0	A_1	A_2	A_3
0.45	0.33	0.19	0.03

9 Hack Lab 2 *Assigned Passwords in the Clear*

Throughout this book you will discover a number—in particular, four—of what we call Hack Labs. These labs are designed to give students practical experience in dealing with a number of cybersecurity issues that are of critical concern in the protection of computing environments.

The other purpose for these labs is that it is not necessary, but there could be a supportive physical computer lab to carry out these projects. They can also be done on a student's own computing equipment and do not have to be done within a fixed lab period.

When these have been offered by the authors, they have usually allowed the students a week to carry out the research and submit the results.

9.1 HACK LAB 2: ASSIGNED PASSWORDS IN THE CLEAR

This Hack Lab deals with ways to discourage online companies from requiring a user to create a password to avail himself or herself of the company services. Most of us who are frequent users of online services find that they may have to manage many such passwords. However, a number of companies, presumably with little concern for security, have the habit of requiring the user to obtain a password by signing up and then having it emailed to the user's account and displaying the host-created password "in the clear."

Instructions:

a. Find a (or many) website(s) for which you must create a password, and then the password is sent to you in email *in the clear*!
b. No responses categorized in plaintextoffenders.com will be accepted. Deadline: 1 week after lab announcement, but this site may give you some good ideas.

REFERENCE

Plaintext Offenders. 2018. http://plaintextoffenders.com/ password in clear plaintext offenders.com

PROBLEM

9.1 How would you prevent exposure of your password if it is delivered from a website in the clear?

10 Origins of Cryptography

All of the preceding sections have led us to what is perhaps the most important component of creating secure environments. This is the study and application of methods to transform information by a secret function so that it will be extremely difficult for an attacker to be able to interpret the message in question.

The approach that is central to this and any course on cybersecurity is known as cryptology, which is in effect a combined term stemming from the study of cryptography (from the Greek κρυπτο [krypto or *hidden*] and γραφια [grafia or *writing*]). The second component of cryptology is the field of cryptanalysis, or the study of ways of breaking cryptographic methods. Both fields have an extremely long history far predating the computer era. In fact, one of the earliest cryptographic methods is said to have been invented by Julius Caesar over 2000 years ago. The so-called "Caesar shift" is not a practical cryptographic method in these times, but it is useful to describe, primarily to set the terms of reference for a cryptographic system (Patterson, 1987).

10.1 CAESAR SHIFT

In the Caesar shift, as well as in any cryptographic method or algorithm, we have a text that is required to be hidden, called the plaintext. We also have a number of different possible ways of transforming that text, each one of those ways determined by the choice of the transformation that we call the key. Once the transformation determined by the key is applied to the plaintext, the result is referred to as the encrypted text or the ciphertext.

In the example of the Caesar shift, suppose that we wish to transmit the message: "ROME WAS NOT BUILT IN A DAY". And, for this transmission today, we choose as the key the number 5, by which we mean that each letter in the message will be advanced by five positions in the alphabet—in the case of letters near the end of the alphabet, if we advance beyond the letter Z, we continue back to the beginning or A. Thus, our plaintext message becomes:

```
ROME WAS NOT BUILT IN A DAY
SPNF XBT OPU CVJMU JO B EBZ
TQOG YCU PQV DWKNV KP C FCA
URPH ZDV QRW EXLOW LQ D GDB
VSQI AEW RSX FYMPX MR E HEC
WTRJ BFX STY GZNQY NS F IFD
```

so, the ciphertext to be transmitted to the receiver is "WTRJ BFX STY GZNQY NS F IFD".

In order for any cryptographic method to be useful, there must be an inverse to the key in order to allow the receiver to apply that inverse key and thus retrieve the original message. In the case of the Caesar shift, the inverse key is the inverse of the value of the key in arithmetic modulo 26—or, in other words, the number that when added to the key (5) adds to 26, in other words $5 + 21 \equiv 0 \pmod{26}$.

Thus, the method of decrypting the encrypted text is to use the same process, in this case by advancing each letter 21 times. Of course, the equivalent is to back each letter in the encrypted text up 5 times. Again, we use the "cyclic" alphabet: after Z we go to A.

In order to use this or any other cryptographic method, we require a method for formulating plaintext; a number of possible transformations, each defined by a key; and a system or alphabet for the ciphertext. Finally, we require that the choice of the key or transformation be kept secret, but somehow it must also be shared between the sender and the receiver of the encrypted message so that the receiver can apply the inverse key in order to retrieve the original plaintext.

In the case of the Caesar shift, one failing is that there are a maximum of 26 possible keys, one for each letter of the alphabet, so for the attacker trying to break the Caesar shift, it is only necessary to try all 26 potential keys, and one of them will produce the desired result.

We will describe a few other historic methods used in the paper-and-pencil era, only to illustrate how the complexity of these methods has evolved over the centuries.

10.2 SUBSTITUTION AND TRANSPOSITION

Most cryptographic methods over time have relied on two philosophically different approaches to the transformation of information. These are called substitution and transposition. The Caesar example cited above uses the principle of substitution; in other words, each character in the plaintext is transformed according to a function of the character symbol itself. In the case of our example above with the key "5", the 26 letters of the alphabet are transformed as follows: A → F, B → G, C → H, …, Z → E.

The other approach, transposition, which we will see later, performs the transformation on the position of the symbol in the plaintext list. A very simple example might be described as (1 2 3 4) → (3 4 2 1), by which we mean that the first of a four-letter group moves to the third position, the second to the fourth, and so on. Thus, in this case, the message "EACH GOOD DEED PAYS" becomes "HCEA DOGO DEDE SYPA". The inverse key is therefore (1 2 3 4) → (4 3 1 2).

10.3 THE KEYWORD MIXED ALPHABET CRYPTOSYSTEM

The keyword mixed alphabet uses for its keys the set of all words, with duplicate letters removed, in the English language. Indeed, the requirement for words to be English words (in addition to being xenophobic) is imposed only because the distribution of the keyword is much simpler if it is a word rather than an arbitrary character string.

> The messenger, having ridden from Lexington to Valley Forge in half a day, was exhausted, out of breath, and indeed near death as he approached General Washington to tell him the secret key for the cipher system: "XRUTGDKWQFP", he panted, then expired. Did he say "XRUTGDKWQFP" or "XRUTGDKWQFT"? puzzled General Washington.

The method itself uses the keyword to define a mapping or permutation of the message space alphabet (Σ, the symbols of the English language alphabet—52 characters including upper and lowercase letters). To define the key, the alphabet is written in its normal order, and under it is written a permuted alphabet, with the letters of the keyword followed by the remaining letters of the alphabet. For the moment, consider only that we are using just the 26 uppercase characters.

Suppose the keyword is: FACETIOUSLY
Then the regular alphabet is (with the spaces inserted just for readability):
ABCDE FGHIJ KLMNO PQRST UVWXY Z
and the permuted alphabet is:
FACET IOUSL YBDGH JKMNP QRVWX Z
Written for ease of reading, we have:

```
ABCDE FGHIJ KLMNO PQRST UVWXY Z
FACET IOUSL YBDGH JKMNP QRVWX Z
```

The encryption maps each letter of the message text to a letter of cipher text according to the permutation defined above. Thus, "MAY THE FORCE BE WITH YOU" becomes "DFX PUT IHMCT AT VSPU XHQ", or, more likely, "DFXPU TIHMC TATVS PUXHQ". It is common practice in older cryptosystems to group the text in equal size blocks. On the one hand, the spaces allow for easier human reading; but more importantly, the normal position of the blanks or spaces tells us the word lengths.

10.4 THE VIGENÈRE CRYPTOSYSTEM

The Vigenère cipher was a widely used cryptosystem dating back to the sixteenth century, using a keyword combined with a Caesar shift. If the keyword is "FACETIOUSLY", an 11-letter word; as before, the encryption will use 11 different Caesar shifts periodically. (Each letter determines a Caesar shift, or modular addition.) Suppose that 0 ↔ A, 1 ↔ B, …, 25 ↔ Z, as usual. Then, the first letter to be encoded uses the shift corresponding to F, the second to A, the third to C, and so on until the cycle repeats:

Choose a key word, perhaps:
Plain text:

```
IT'S A LONG WAY TO TIPPERARY …
```

Key:

```
FA C E TIOU SLY FA CETIOUSLY
```

Cipher text:

```
OUV F FXCB PMX ZP WNIYTMTDX
```

10.5 ONE-TIME PAD ENCRYPTION

The "one-time pad" plays an important role in the history of encryption. It was first described in 1882 by Frank Miller, and U.S. Patent 1,310,719 was issued to Gilbert Vernam for a physical device based on this encryption method (Patterson, 1987).

The principle of the one-time pad is an elaboration on the Vigenère cryptosystem as described above. In the Vigenère system, the key word is chosen (FACETIOUSLY in our example), and then encryption consists of a different Caesar shift based on the corresponding element in the pad corresponding to the plaintext letter in the same position. However, the breaking or cryptanalysis of the Vigenère method is based on the realization that after a certain number of plaintext letters are encrypted, the period related to the key may be determined. In the case of the example with keyword FACETIOUSLY, the same Caesar shift will take place with characters in positions 1, 12, 23, 34 and so on: in other words, every 11th letter.

The concept with the one-time pad is that the length of the key word can be stretched to an enormous length. Indeed, for the true one-time pad, the keyword should be of infinite length. However, the creation of an infinite-length string that does not repeat is practically impossible. But, it is possible using randomization techniques to generate long sequences of letters that may only repeat after thousands or indeed millions of terms in a sequence.

Thus, it is possible to create a "pad," that is, an extremely long keyword, so that there is no repetitive pattern as long as the number of characters in the plaintext to transmit is smaller than the length of the key word.

The problem, however, is that it becomes extremely difficult to manage a one-time pad with an extremely long keyword. Among the problems: (1) if a person is to memorize the one-time pad of, let us say, 1 million characters, our human memory is insufficient to be able to maintain this; (2) if we attempt to maintain the pad by writing it down, we have exposed the system to potential theft of the long and written key; and (3) even when the key is successfully passed on to the potential receiver, exposure of the long word constitutes a third problem.

As a consequence, the practical use of a one-time pad system depends entirely on an environment for encryption where both the sender's ability to encrypt and the receiver's ability to decrypt are carefully controlled. Thus, for all practical purposes, a one-time pad is only useful in controlled military environments. In most Hollywood depictions of a military coding system where the receiver must decrypt the message, he or she must go to a locked safe, usually worked by two persons who simultaneously ensure that both have the same "codebook," and the operator reads the keyboard one-time pad from the codebook. And again, to be useful, the length of the codebook must be greater than the length of all messages so distributed.

Long after one-time pads began to be used in military applications, it was proved by Claude Shannon of Bell Laboratories in 1949 (Shannon, 1949) that the one-time pad could be proved in its original description as the most secure encryption possible, despite the fact that the true one-time pad of infinite length is essentially impossible to construct.

For example, the pad is:

S	I	S	A	V	C	U	I	I	D	F	Z	D	M	M
M	Y	G	L	X	S	K	G	K	S	I	J	C	A	H
O	O	F	N	C	W	F	L	V	L	Z	R	M	G	Y
V	G	I	E	D	Q	A	O	X	J	K	I	Y	W	G
E	N	D	I	D	V	W	P	Y	D	V	X	E	R	W
Y	U	X	I	I	W	L	R	V	J	C	I	Q	U	U

L	U	Z	Q	I	A	Y	D	S	I	U	A	M	A	B
O	D	A	Y	O	P	P	G	R	V	Z	Z	U	C	G
I	K	N	F	P	O	F	P	B	I	R	W	P	C	C
U	D	O	H	Q	D	Z	D	B	O	W	B	Q	G	H
F	C	J	E	I	I	R	R	V	G	N	Y	Z	F	W
X	G	U	I	K	D	W	X	P						

It is constructed by calculating 174 consecutive results of choosing a number at random (in this case 8683, between 1 and 17,388 (17,389 is a prime number—this is critical), then successively multiplying the number by 8683 and taking the remainder with respect to 17,389. We can generate 17,388 such numbers distinctly. Then we take the remainders so generated with respect to 26 (giving 26 distinct values), which we assign in the same fashion to the letters of the alphabet.

All of the numbers in the preceding paragraph could be chosen in many ways. The 17,389 can be replaced by any large-enough prime number p, the 8683 (=n) at random in the range $1 < n < p$, and the 174 just an example; it could be any number less than 174.

So, let's use this quote from Albert Einstein to encrypt:

The difference between stupidity and genius is that genius has its limits.

THE DIFFERENCE BETWEEN STUPIDITY AND GENIUS IS THAT GENIUS HAS ITS LIMITS

```
THEDI  FFERE  NCEBE  TWEEN  STUPI  DITYA  NDGEN  IUSIS  THATG
ENIUS  HASIT  SLIMI  TS

SISAV  CUIID  FZDMM  MYGLX  SKGKS  IJCAH  OOFNC  WFLVL  ZRMGY
VGIED  QAOXJ  KIYWG  EN
DIDVW  PYDVX (etc. --- not needed for this encryption)

MQXEE  IANAI  TFIOR  GVLQY  LEBAB  MSWZI  CSMSQ  FAEEE  TZNAF
AURZW  YBHGD  DUHJP  YG
```

Each character in the encryption is computed by the corresponding Caesar shift, for example:

Using the correspondence $A \leftrightarrow 1, B \leftrightarrow 2, C \leftrightarrow 3, \ldots, Z \leftrightarrow 26$, add the value for each plaintext character to the value of the corresponding element of the pad (in mod 26, or subtracting 26 if the sum is >26. The encryption of these characters is the character corresponding to the sum.

```
T → 20, S → 19; 20 + 19 = 39; remainder mod 26 = 13, 13 → M
H →  8, I →  9;  8 +  9 = 17; remainder mod 26 = 17, 17 → Q
E →  5, S → 19;  5 + 19 = 24; remainder mod 26 = 24, 24 → X
```

And so on.

10.6 THE PLAYFAIR SQUARE

Here is a more sophisticated method from the nineteenth century known as the Playfair Square (attributed to the British scientist Charles Wheatstone in 1854 but bearing the name of Lord Playfair for marketing purposes). To create a Playfair Square, we will build a 5×5 square using the regular English alphabet (Patterson, 1987).

Whoops! The alphabet has 26 letters and yet $5 \times 5 = 25$.

Our task is to reduce the alphabet to 25 letters, which we accomplish by merging two letters together, I and J. So, in the algorithm, in the plaintext, every J is changed to an I.

Now pick a word to form the key. If the keyword has a duplicate, for example, suppose the keyword is CONSPIRATORIALLY, then by deleting, we will use CONSPIRATLY as the key.

Now create the Playfair Square by writing the letters of the key into the 5×5 square in row order. Then fill in the rest of the (25-letter) alphabet, also in row order. The Playfair Square encryption method is based not on substituting single letters (therefore 26 different substitutions) when taking letters in pairs (or digrams). There are $26 \times 25 = 650$ digrams; therefore, the total number of possible permutations of those 650 symbols is a number approximately 8.1×10^{1547}, that is, greater than a 1 followed by 1547 zeros. In this case, the Playfair Square is:

```
C O N S P
I R A T L
Y B D E F
G H K M Q
U V W X Z
```

The 5×5 square is essentially the key, although all that has to be shared with the receiver is the keyword, CONSPIRATORIALLY in this case (Figure 10.1).

Now to encrypt a message. The square is based on the symbols being not letters but pairs of letters. We first break the plaintext message into digrams. For example, the plaintext message COMMITTEE becomes:

CO MM IT TE EZ (if the plaintext has an odd number of letters, just attach any letter at the end).

Encrypt Rule 1: if the digram is in the same row or column, shift each character to the right (for a row) or down (for a column), thus: CO → ON, IT → RL

Encrypt Rule 2: if the digram defines a rectangle, take the other corners of the rectangle, in the order of the row of the first message character. Thus, for example, EZ → FX

But, for our keyword CO MM IT TE EZ, how do we deal with MM? The answer is, we don't. We preprocess the message by inserting a letter (Q is best) between

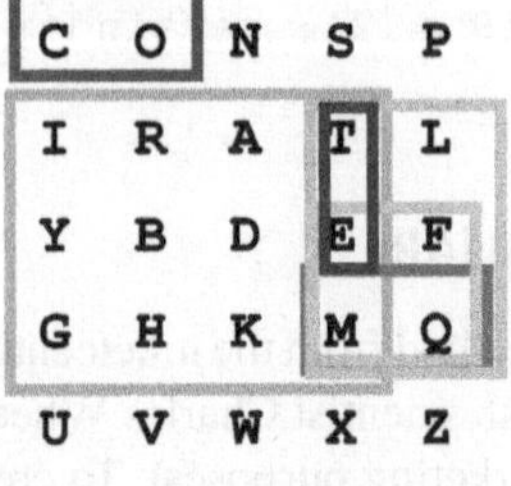

FIGURE 10.1 The Playfair Square.

every pair. Thus actually COMMITTEE becomes COMQMITQTEQE or CO MQ MI TQ TE QE. Thus, we have ensured that in the text to be encrypted, we will have no double letters and thus either Rule 1 or Rule 2 will apply. The encryption of the new plaintext becomes

- □ CO → ON
- □ MQ → QG
- □ MI → GT
- □ TQ → LM
- □ TE → EM
- □ QE → MF

Now you can send the message. The receiver has the same square and has two encrypt rules, Rule 1 going left if the pair is in the same row, going up if they're in the same column, and the rectangle Rule 2 is the same as for encryption.

Finally, as the receiver has the decrypted but slightly altered message, the receiver pulls out all the Qs inside pairs, and interprets "I" as "J" where necessary.

When we change all Js into Is, this will almost never confuse issues in the decryption. And, with respect to the inserted Qs between double letter pairs, can this ever become a problem? Since in English, a Q is almost always followed by a U, this will almost never create a problem—except possibly for VACUUM.

10.7 ROTOR MACHINES

Early in the twentieth century, machine production of ciphers became possible. To mechanize the production of ciphertext, various devices were invented to speed the process. All important families of such devices were rotor machines, developed shortly after the time of World War I, to implement Vigenère-type ciphers with very long periods.

A rotor machine has a keyboard and a series of rotors. A rotor is a rotating wheel with 26 positions—not unlike the workings of an odometer in an automobile. Each position of the rotor wheel completes an electric contact, and depending upon the position, determines a different Caesar shift. When a key on the keyboard is depressed, a letter is generated dependent on the position of the rotors.

10.8 WORLD WAR II AND THE ENIGMA MACHINE

A United States patent for a rotor machine was issued in 1923 to Arthur Scherbius (IT History, 2018). It essentially formed the basis for the German Enigma machine. A variation on the rotor machine design was developed by the German Armed Forces leading up to World War II and called the Enigma machine. The now well-documented British success at breaking the code at Bletchley Park was a major factor in the war effort. In 2014, the movie *The Imitation Game* depicted reasonably well the efforts led by Alan Turing in breaking the Enigma code (Sony, 2014). Turing, considered by many the father of computer science, died tragically in 1954 at age 41 (Figures 10.2 and 10.3).

FIGURE 10.2 The Enigma machine.

FIGURE 10.3 Alan Turing.

REFERENCES

IT History Society. 2018. Dr. Arthur Scherbius Bio/Description. https://www.ithistory.org/honor-roll/dr-arthur-scherbius
Patterson, W. 1987. *Mathematical Cryptology.* Rowman and Littlefield, 318 pp.
Shannon, C. 1949. Communication theory of secrecy systems. *Bell System Technical Journal,* 28(4), 656–715.
Sony Pictures Releasing. 2014. *The Imitation Game* (film).

PROBLEMS

10.1 Find two strings of the greatest length that make sense in English and are related by a Caesar shift. [Example: t4(CAP) = GET. Thus, t1(BZO) = CAP and t5(BZO) = GET. Ignore blanks.] If the "CAP/GET" example is length 3, the record in our classes is length 7.

10.2 What is unique about the keyword FACETIOUSLY?

10.3 Decrypt the following cipher using Caesar's shift:

```
RVPXR  WPCXV  JTNQR  VJWXO  ONAQN  LJWCA  NODBN
```

10.4 Using the (1 2 3 4 5) → (4 3 5 1 2), encrypt "FOURSCORE AND SEVEN YEARS AGO".

10.5 Decrypt this message that uses a transposition cipher:

```
RTEH IANI ANPS SIIN NMIA NLIY PTEH NLIA
```

10.6 Using the Vigenère cipher method with a 10-letter keyword of your choice, encrypt the message:

I have a dream that one day on the red hills of Georgia the sons of former slaves and the sons of former slave owners will be able to sit down together at the table of brotherhood.

10.7 Create a Playfair Square with key word IMPOSSIBLE and encrypt the message JUSTICE EVENTUALLY WINS OUT.

10.8 Billary and Hill use a Playfair cipher to prevent their communications from being intercepted by an intruder, Trumpy. They have agreed on a common key, from the word "ALGEBRAICALLY".

a. Construct their jointly used Playfair Square.

A	L	G	E	B
R	I	C	Y	D
F	H	K	M	N
O	P	Q	S	T
U	V	W	X	Z

b. Billary receives the following Playfair-encrypted message from Hill:

```
RP  GX  HI  PG  PG  BS  SY  RT  RV  TO  SO  ML  AT
SP  QW  MG  SG  VH  KB
```

Find Billary's decryption.

11 Hack Lab 3 *Sweeney Method*

Throughout this book you will discover a number—in particular, four—of what we call Hack Labs. These labs are designed to give students practical experience in dealing with a number of cybersecurity issues that are of critical concern in the protection of computing environments.

The other purpose for these labs is that it is not necessary, but there could be a supportive physical computer lab to carry out these projects. They can also be done on a student's own computing equipment and do not have to be done within a fixed lab period.

When these have been offered by the authors, they have usually allowed the students a week to carry out the research and submit the results.

11.1 HACK LAB 3: SWEENEY PRIVACY STUDY

Dr. LaTanya Sweeney, now professor of government and technology at Harvard, published in her PhD thesis in computer science at MIT that it was possible to identify uniquely over 85% of the United States population with only three pieces of easily available information: the gender, the exact birth date, including year, of the individual, and the postal zip code of their address (Sweeney 2000) (Figure 11.1).

The purpose of this hack is to confirm the research of Dr. Sweeney's research on privacy. Her research showed definitively that 87% of the U.S. population can be uniquely identified by three commonly available data points: { birthdate of the individual including year, gender, and U.S. zip code (5-digit version) of residence }.

A test bed of 20 data sets is provided as samples for carrying out this lab. The objective for students is to try to identify the individuals for whom, in most cases, sufficient information is provided to determine an individual in the United States who fits the given data regarding birth date and zip code of residence.

When this experiment was conducted by our students at Howard University, on separate occasions, usually about half the class would identify 10 individuals given their personal data within 24 hours.

The only additional advice given students was that the subjects to find had some measure of celebrity.

With this information, and with general access to health records without names being protected or encrypted in a health database, a health researcher could construct a legitimate query that would result in only one record being returned, and thus by independently discovering the information described above (exact birth date and zip code), discover without hacking or illegal access the health records of an individual. This demonstrates a problem in maintaining secure databases.

FIGURE 11.1 Professor LaTanya Sweeney speaking on the social security number crisis at Harry Lewis's class Life, Liberty, and Happiness after the Digital Explosion, October 20, 2008. From the Creative Commons Attribution 2.0 Generic license. (From Flickr, cropped. Author: "arcticpenguin.")

In fact, Dr. Sweeney was able to discover that the then governor of Massachusetts had not disclosed health information that could have affected his performance in public office as governor.

In this assignment, I have provided the exact birth dates and zip codes of 10 real people, using publicly available information, and constructed (with *fake* diseases) the results of a query or queries into a health database that returned the following (names being obscured or suppressed):

Name	Birth Month	Birth Day	Birth Year	Zip Code	Disease
	November	14	1954	94306	Kidney disease
	March	1	1994	90068	Pneumonia
	June	8	1925	04046	Arthritis
	February	20	1963	85258	HIV
	June	1	1937	38921	Migraine headaches
	August	30	1930	68132	Ruptured spleen
	July	31	1958	75220	Alzheimer's disease
	August	11	1965	91344	Gout
	March	25	1942	48302	Chronic bronchitis
	October	28	1955	98112	Macular degeneration

(*Continued*)

Name	Birth Month	Birth Day	Birth Year	Zip Code	Disease
	September	24	1981	33154	Kidney disease
	December	30	1984	33133	Pneumonia
	February	11	1964	99654	Arthritis
	May	6	1961	91604	HIV
	June	14	1946	20500	Migraine headaches
	April	25	1940	10964	Ruptured spleen
	January	11	1971	07458	Alzheimer's disease
	June	8	1958	91356	Gout
	February	12	1990	20105	Chronic bronchitis
	July	3	1962	90210	Macular degeneration

Your assignment is to use this information to discover the individual in each case. Like any good hacker, you will not discuss your work with anyone else—after all, the person you might discuss this with might turn you in to authorities. For each name you discover, email that person's name and the disease to me by email. You get 1 point (out of 10) for each correct solution.

Except: beginning tomorrow at 11:59 PM, I will reveal in an email one of the persons involved. I will reveal one more name and disease each 24 hours afterward. Obviously, you won't get a point for a submission after I've revealed the name. Thus, the lab will end in 10 days.

REFERENCE

Sweeney, L. 2000. *Simple Demographics Often Identify People Uniquely.* Data Privacy Working Paper 3. Pittsburgh: Carnegie Mellon University.

PROBLEMS

11.1 Carry out this Hack Lab, first for the first table above, then for the second table. You will be above the 50th percentile of our students if you complete either table within 1 day.

11.2 Create your own set of 10 dates and zip codes, and exchange with a classmate.

11.3 Give an estimate of Sweeney's figure of 87% individual identification for:

a. Canada
b. United Kingdom
c. Brazil

12 Hacker Personalities
Case Studies

In studying various persons who have spoken or written about their exploits as hackers or cybersecurity criminals, it seems that we can gain some knowledge about their technical achievements by analyzing their personalities.

One person who has been well known in the hacker community for many years is Kevin Mitnick. Mitnick enjoyed great success in compromising telephone networks (phone phreaking) until he went to federal prison in the 1980s. Subsequently, he has become an influential writer and speaker about cybersecurity. One of his best-known books is *The Art of Intrusion: The Real Stories Behind the Exploits of Hackers, Intruders and Deceivers*, written with William Simon (Mitnick and Simon, 2005).

This book details the exploits of numerous hackers, alleged to be true stories, and in many of the cases describes enough about the interests and motivations of the subjects to give us some insight into their personalities.

We select just a few examples from *The Art of Intrusion* to identify some of the personalities and what drove them to the exploits described in the book.

12.1 COMRADE

The hacker known as Comrade began his exploits as a teenager living in Miami. About some of his early works, Comrade said, "we were breaking into government sites for fun." Comrade developed a friendship with another hacker with an Internet name of neoh, another young man who was only a year older than Comrade, but who lived 3000 miles away. About his interests, neoh said, "I don't know why I kept doing it. Compulsive nature? Money hungry? Thirst for power? I can name a number of possibilities." Also, neoh, in corresponding with author Mitnick, wrote: "You inspired me… I read every possible thing about what you did. I wanted to be a celebrity just like you."

Another person, named Khalid Ibrahim, who claimed to be from Pakistan, began to recruit Comrade and neoh. Khalid's interest was in working with other hackers who might be willing to hack into specific targets—first in China, then in the United States. Khalid indicated that he would pay cash for successful penetration into the targets he indicated.

Comrade's interest, as he indicated, was that he knew that Khalid "was paying people but I never wanted to give up my information in order to receive money. I figured that what I was doing was just looking around, but if I started receiving money, it would make me a real criminal."

12.2 ADRIAN LAMO

Adrian Lamo, as a teenager, lived in New England, and developed his hacking skills at an early age.

Mr. Lamo, according to his parents, was involved in hacking because of a number of specific well-known hackers who were his inspiration. His strategy in hacking was to understand the thought processes of the person who designed the subject of his attacks, a specific program or network. In one case, he discovered a customer who asked for assistance with stolen credit card numbers, and the technicians who were supposed to assist didn't bother responding. Then Adrian called the victim at home and asked if he had ever gotten a response. When the man said no, Adrian forwarded the correct answer and all the relevant documentation regarding the problem. As Lamo said, "I got a sense of satisfaction out of that because I want to believe in a universe where something so improbable as having your database stolen by somebody… can be explained a year later by an intruder who has compromised the company you first trusted."

Adrian's description of his philosophy can be summarized as: "I believe there are commonalities to any complex system, be it a computer or the universe…. Hacking has always been for me less about technology and more about religion."

12.3 GABRIEL

Gabriel lives in a small town in Canada, and his native language is French. Although he sees himself as a white-hat hacker, he occasionally commits a malicious act when he finds a site "where security is so shoddy someone needed to be taught a lesson." As a young man, he found details about the IP addresses of a small bank in the U.S. south that nevertheless had extensive national and international ties. He discovered that one of the bank's servers ran software that allows a user to remotely access a workstation. Eventually he found ways to remotely access terminal service, so he could essentially own the potential system. He also found the password for the bank's firewall, so hacking into that one machine gave him access to other computer systems on the same network. As a consequence, Gabriel had access to great deal of internal information, but he did not have any interest in stealing funds. "I found a lot of documents on their server about physical security, but none of it was related to hackers… they're doing a good job on physical security, but not enough for computer security."

12.4 ERIC

Eric is a security consultant who complains that "when I report a vulnerability, I often hear, 'it's nothing. What's the big deal?'" Eric and some colleagues of his admit to collecting different types of server software and reached the point where they owned the source code of all the major products in the category with only a single exception. Eric said, "this was the last one I didn't have, and I don't know why, it was just interesting to me to break into that one." Eric persisted in this quest for well over a year, and eventually "I finally found the CEO's computer and that was kind of interesting. I port-scanned it for a couple of days and there would be just no response, but I knew his computer was there."

It turns out the CEO used his laptop and was only on for about 2 hours every morning. He would come into his office, check his email, and then leave. Eric felt challenged enough to try to determine the CEO's password: "You know, the truth is, I can't explain it. It's just an ability I have to guess the passwords people use. I can also know what sort of passwords they would use in the future. I just have a sense for that. I can feel it. It's like I become them and say what password I would use next if I was them."

12.5 WHURLEY

The last example chosen from Mitnick's book involves a security consultant whose pseudonym is "Whurley." He was hired by a resort in Las Vegas to perform a number of security audits. He noted that the employees who were supposed to provide security were very lax, including an employee would lose his badge all the time, or just share it with another employee to get in for free meals. "As I'm walking out, I see an open empty office. It has two network ports, but I can't tell if they're hot by just looking at them, so I go back to where the assistant is sitting and tell her that I forgot I was supposed to look at her system and the one in 'the boss's office.' She graciously agrees and lets me sit at her desk. She gives me her password when I asked, and then has to use the restroom. So, I tell her I'm going to add a "network security monitor" and show her the wireless access point. She replies, 'whatever. I really don't know much about that geeky stuff.'"

In addition to relating individual hacker personalities, a number of authors have attempted to describe generic personality traits of hackers.

12.6 HACKER PERSONALITY DESCRIPTIONS

Lee Munson is a security researcher for Comparitech and a contributor to the Sophos Naked Security blog. Munson wrote (2016):

> It's hard to pin down just a few personality traits that define a hacker. A typical hacker profile is a male, age 14 to 40, with above-average intelligence, obsessively inquisitive with regards to technology, non-conformist, introverted, and with broad intellectual interests. A hacker is driven to learn everything he can about any subject that interests him.
>
> In fact, most hackers that excel with technology also have proficiency in no technological hobbies or interests. Hackers tend to devour information, hoarding it away for some future time when a random bit of technical trivia may help them solve an intriguing problem. Hackers seem especially fond of complex intellectual challenges and will move on to a new project once the challenge and novelty wears off…
>
> Financial gain motivates some crackers. Credit card and bank fraud present opportunities to use cracking to increase personal wealth.

Eric Stephen Raymond is the cofounder of the Open Source Initiative, an organization that builds bridges between the hacker community and business. Raymond wrote (2015):

> Although high general intelligence is common among hackers, it is not the *sine qua non* one might expect. Another trait is probably even more important: the ability to

mentally absorb, retain, and reference large amounts of "meaningless" detail, trusting to later experience to give it context and meaning. A person of merely average analytical intelligence who has this trait can become an effective hacker, but a creative genius who lacks it will swiftly find himself outdistanced by people who routinely upload the contents of thick reference manuals into their brains. [During the production of the first book version of this document, for example, I learned most of the rather complex typesetting language TeX over about four working days, mainly by inhaling Knuth's 477-page manual. My editor's flabbergasted reaction to this genuinely surprised me, because years of associating with hackers have conditioned me to consider such performances routine and to be expected.—ESR]

Contrary to stereotype, hackers are not usually intellectually narrow; they tend to be interested in any subject that can provide mental stimulation, and can often discourse knowledgeably and even interestingly on any number of obscure subjects—if you can get them to talk at all, as opposed to, say, going back to their hacking…

In terms of Myers-Briggs and equivalent psychometric systems, hackerdom appears to concentrate the relatively rare INTJ and INTP types; that is, introverted, intuitive, and thinker types (as opposed to the extroverted-sensate personalities that predominate in the mainstream culture). ENT[JP] types are also concentrated among hackers but are in a minority.

Rick Nauert has over 25 years of experience in clinical, administrative, and academic healthcare. He is currently an associate professor for the Rocky Mountain University of Health Professionals doctoral program in health promotion and wellness. And Nauert likens the personality traits of hackers to the symptoms of autism (2016).

Online hacking costs the private and corporate sectors more than $575 billion annually. While security agencies seek out "ethical" hackers to help combat such attacks, little is known about the personality traits that lead people to pursue and excel at hacking.

New research shows that a characteristic called systemizing provides insight into what makes and motivates a hacker. Intriguingly, the personality traits are similar to many autistic behaviors and characteristics.

"We found a positive association between an individual's drive to build and understand systems—called systemizing—and hacking skills and expertise," said Dr. Elena Rusconi of the Division of Psychology at Abertay University in Dundee, U.K.

"In particular, we found that this drive is positively and specifically correlated with code-breaking performance."

What is systemizing? Systemizing is the preference to apply systematic reasoning and abstract thought to things or experiences. It is theorized to exist on a continuum with a personality trait called empathizing, a preference for being agreeable and able to empathize with others. The preference for systemizing is frequently associated with autism or Asperger's, a milder form of autism.

In the study, Rusconi's group found that volunteer "ethical" hackers performed far above average on a series of code-breaking challenges designed to assess their systemizing skills.

REFERENCES

Mitnick, K. and Simon, W. 2005. *The Art of Intrusion: The Real Stories Behind the Exploits of Hackers, Intruders and Deceivers*. Indianapolis: Wiley Publishing.

Munson, L. 2016. Security-FAQs, http://www.security-faqs.com/what-makes-a-hacker-hack-and-a-cracker-crack.html

Nauert, R. 2016. PsychCentral.com, https://psychcentral.com/news/2016/06/02/some-personality-traits-of-hackers-resemble-autism/104138.html

Raymond, E. S. 2015. Catb.org, http://www.catb.org/jargon/html/appendixb.html

PROBLEMS

12.1 Read *The Art of Intrusion*. Identify any of the characters portrayed as female, or any of the characters described (or that you would estimate) as being over 50 years in age.

12.2 Why did Mitnick, the coauthor of *The Art of Intrusion*, go to jail?

12.3 Would any of the cases in *The Art of Intrusion* be described as cases of social engineering?

12.4 Critique the description Lee Munson has provided in this chapter of the personality traits of a hacker.

12.5 Research the Myers-Briggs personality types indicator system (see the chapter on personality tests, also https://upload.wikimedia.org/wikipedia/commons/1/1f/MyersBriggsTypes.png). Identify categories unlikely to be attributed to a hacker.

12.6 Discover if there are any professions that seem to have an overabundance of persons with Asperger's syndrome.

13 Game Theory

An important technique in the area of cybersecurity can be in the application of what we know as game theory (Morris, 1994; Von Neumann and Morgenstern, 1944).

For the purposes of this course, we will develop just an introductory exposition to game theory, notably what is now referred to as "two-person, zero-sum games." The term *zero-sum* refers to the constraint that whatever one party gains in such a game, his or her opponent loses. Perhaps the simplest example of such a two-person zero-sum game involves the toss of a coin. The sides of the coin are labeled heads or tails, and one player tosses while the other guesses the outcome of the toss. Let's call the players Mary and Norman, and the results of the toss either H or T. Mary tosses the coin, and while it is in the air, Norman guesses the outcome. Clearly, there are four potential results in this game. In two-person game theory, we describe such a game by a matrix, with the rows of the matrix representing the choices for the first player and the columns the choices for the second. In this game, there are only two choices for each player, so we establish a 2×2 matrix.

13.1 PAYOFF

The "payoff" or outcomes are represented by values in the body of the 2×2 matrix. If we assume that if the first player guesses the result of the second player's toss, he or she wins one unit of currency, say one dollar, then the entire game can be described as in this matrix (Table 13.1).

By convention, we always label the row player's positive outcome with a positive number; the column player interprets that value as a loss. So, in the example above, if Mary tosses heads and Norman guesses tails, then Mary wins a dollar. We enter a -1 in the T row for Norman—which Mary interprets as a $+1$ for her outcome.

More generally speaking, in such a game, each player knows in advance this payoff matrix, but game theory provides the analytical approach to enable each player to determine his or her best strategy.

In this example, the coin tosser doesn't have a strategy, but we could alter the game very slightly by saying that rather than conducting a toss, Mary would simply secretly choose one of her outcomes (H or T), and Norman would guess at the result, and then they would compare their choices.

In this model, each would develop a strategy to determine their choice, but a quick analysis would show that there is no winning strategy, as each player as a 50–50 chance of winning.

Suppose, however, that we create a game where the winning strategies might be different for each player. For example, with two players, each will call out either the number one or two. One player is designated as "odd"—let's say he is Oliver—the other player is "even," and she will be known as Evelyn. If the result of the sum of each player's call is odd, Oliver wins that amount; if the result of the sum is even,

TABLE 13.1
Heads or Tails

	Norman (Column Player)		
	Choice	**H**	**T**
Mary (row player)	H	+1	−1
	T	−1	+1

TABLE 13.2
Calling "One" or "Two"

	Evelyn (Even)		
	Choice	**One**	**Two**
Oliver (odd)	One	−2	3
	Two	3	−4

Evelyn wins that amount. Using the model described above for the game matrix, we have Table 13.2.

Rather than thinking of this or any of our games as a one-time event, consider what players would do if the game may be repeated. Each tries to develop a strategy whereby they can improve their chances. One "strategy" might be simply a random choice of their play, in which case in the above example, Oliver would choose "one" 50% of the time, and "two" the other 50%. Evelyn would reason similarly.

In such a case, Oliver would lose two dollars half the time and win three dollars the other half if he chose "one"; in other words, $0.5(-2) + 0.5(3) = +0.5$.

On the other hand, if Oliver chose "two" half the time, his result would be $0.5(3) + 0.5(-4) = -0.5$. So, from Oliver's perspective, he should just choose "one" all the time. But Evelyn, being no fool, would discern the strategy, choose "two" all the time, and thus win three dollars at each play. Therefore, in order for a game such as this to be competitive, each player must determine a strategy, or a probability of each potential choice.

Suppose that Oliver chooses "one" 60% of the time (3/5) and "two" 40% of the time (2/5); then this strategy (assuming Evelyn plays randomly) will net him $-2(3/5) + 3(2/5) = 0$ when he calls "one," and when he calls "two," he wins $3(3/5) - 4(2/5) = 1/5$. So, he breaks even over time when he calls "one" but wins 0.5 when he calls "two."

Suppose we let p represent the percentage of times that Oliver calls "one." We would like to choose p so that Oliver wins the same amount no matter what Evelyn calls. Now Oliver's winnings when Evelyn calls "one" are $-2p + 3(1 - p)$, and when Evelyn calls "two," $3p - 4(1 - p)$. His strategy is equal when

$$-2p + 3(1-p) = 3p - 4(1-p),$$

Solving for p gives 7/12, so Oliver should call "one" 7/12 = 58.3% of the time, and "two" 5/12 = 41.7% of the time. Then, on average, Oliver wins −2(7/12) + 3(5/12) = 1/12, or 8.3 cents every time, no matter what Evelyn does. This is called an equalizing strategy.

However, Oliver will only earn this if Evelyn doesn't play properly. If Evelyn uses the same procedure, she can guarantee that her average loss is 3(7/12) − 4(5/12) = 1/12. Because each is using their best strategy, the value 1/12 is called the value of the game (usually denoted V), and the procedure each player uses to gain this return is called the optimal strategy or the "minimax strategy."

What we have just described is a mixed strategy. Where the choices among the pure strategies are made at random, the result is called a mixed strategy. In the game we have just described, the pure strategies "one" and "two" and the mixed strategy lead to the optimal solution with probabilities of 7/12 and 5/12.

There is one subtle assumption here. If a player uses a mixed strategy, he or she is only interested in the average return, not caring about the maximum possible wins or losses. This is a drastic assumption. Here we assume that the player is indifferent between receiving $1 million for sure and receiving $2 million with probability one-half and zero with probability one-half. We justify this assumption arising from what is called utility theory (see Chapter 23 on behavioral economics). The basic premise of utility theory is that one should evaluate a payoff by its utility or usefulness and not its numerical monetary value. A player's utility of money is not likely to be linear in the amount. In a later chapter, we will discuss what is usually called behavioral economics and how it applies to our overall subject.

A two-person zero-sum game is said to be a finite game if both strategy sets are finite sets. Von Neumann has developed a fundamental theorem of game theory, which states that the situation described in the previous holds for all finite two-person zero-sum games. In particular:

The minimax theorem: for every finite two-person zero-sum game, (1) there is a number V, called the value of the game; (2) there is a mixed strategy for Player I such that I's average gain is at least V no matter what Player II plays; and (3) there is a mixed strategy for Player II such that II's average loss is at most V no matter what Player I plays.

If V is zero, we say the game is fair; if V is positive, we say the game favors player I; and if V is negative, we say the game favors player II.

13.2 MATRIX GAMES

To this point, we have only considered two-person games where there are only two pure strategies. Clearly, this is a severe restriction: in general, a player will have many pure strategy options.

More generally, a finite two-person zero-sum game can be described in strategic form as (X, Y, A). In this case, X equals the choice among m pure strategies for Player I, $X = \{x_1, x_2, \ldots, x_m\}$ and Y represents the n pure strategies for Player II, $Y = \{y_1, y_2, \ldots, y_n\}$. With this terminology, we can form the payoff or game matrix with rows and columns corresponding to the choices of each player.

In other words,

$$A = \begin{bmatrix} a_{11} & \cdots & a_{1n} \\ \vdots & \ddots & \vdots \\ a_{m1} & \cdots & a_{mn} \end{bmatrix},$$

and $a_{ij} = A(x_i,y_j)$ = the payoff when Player I chooses strategy x_i and Player II chooses strategy y_j.

A finite two-person zero-sum game in this strategic form (X,Y,A) is sometimes called a matrix game because the payoff function A can be represented by a matrix.

In this form, Player I chooses a row i, Player II chooses a column j, and II pays I the entry in the chosen row and column, a_{ij}. The entries of the matrix are the winnings of the row chooser and losses of the column chooser.

13.3 MIXED STRATEGY

For Player I to have a mixed strategy means that there is an m-tuple p with m distinct probabilities for the choices, $p = \{p_1, p_2, \ldots, p_m\}$, with the additional restriction that the sum of the p_i is 1; that is

$$\sum_{i=1}^{m} p_i = 1.$$

We also denote the strategy for Player II through the n-tuple of probabilities for Player II's choices. Again, the sum of the q_j probabilities must be 1.

The representation by the matrix A is a static description of the game. It merely gives the result if the row player chooses strategy x_i and the column player chooses y_j, as the result or payoff is a_{ij}.

However, using matrix algebra, and considering the range of strategies available to each, the probabilities of choice by the row player can be described by a row vector of strategies $p = [p_1\ p_2 \ldots p_n]$, and the column player's strategies by a column vector $[q_1\ q_2 \cdots q_n]^T$ (where T represents the transpose, simply an easier way of writing the column player's strategies).

The result of the pure strategy can be described as a special case of the mixed strategy by describing in the pure strategy case; the probability vector for the pure strategy is of the form [0 0 … 1 … 0], with the 1 in the ith position if the pure strategy is to choose the ith option, and similarly for the column player.

Given either a pure or a mixed strategy, our objective is to "solve" the game by finding one or more optimal strategies for each player.

As an example, suppose we have a game with three strategies for each player whose payoff matrix is

$$\begin{bmatrix} -1 & 1 & 3 \\ 2 & -1 & -2 \\ 1 & 3 & -1 \end{bmatrix},$$

the row player's strategies are $p = [0.2 \quad 0.4 \quad 0.4]$, and the column player's $q = [0.1 \quad 0.4 \quad 0.5]$ (remember they must add up to 1). Then the average payoff for the row player is pAq.

13.4 SADDLE POINTS

In this most general version of an "m × n" game, we may not be able to determine a solution. One type of such a game that is easy to solve is a matrix game with a "saddle point."

Example: Consider as a game matrix the following:

$$A = \begin{bmatrix} 5 & 3 & 2 & 7 \\ -1 & 0 & 4 & 8 \\ 7 & 5 & 8 & 6 \\ 6 & 4 & -4 & 2 \end{bmatrix}$$

If a game matrix has the property that (1) a_{ij} is the minimum of the ith row, and (2) a_{ij} is the maximum of the jth column, then we call a_{ij} a *saddle point*. By inspecting A above, note that $a_{32} = 5$ is the minimum of the third row, and is also the maximum of the second column. In the case above, the row player can play the third strategy and at least win $5; the column player, playing the second strategy, minimizes his or her losses at the same $5. Thus, in the existence of a saddle point, the value of the game, V, is the value of the saddle point.

Consequently, the strategy vector for the row player is $p = [0 \quad 0 \quad 1 \quad 0]$, and for the column player $q = [0 \quad 1 \quad 0 \quad 0]$.

With simple games as described above, we can check for saddle points by examining each entry of the matrix. However, even in two-person zero-sum games, this can be quite complex. Consider the example of what is usually called "straight poker," where two players receive five cards from a dealer and then bet on their best hand. The number of rows and columns in such a game is 311,875,200.

13.5 SOLUTION OF ALL 2 × 2 GAMES

The general solution of such a game can be found by this two-step strategy: (1) test for a saddle point; (2) if there is a saddle point, it constitutes the solution. If there is no saddle point, solve by finding the equalizing strategy.

Describe the general 2 × 2 game as $A = \begin{bmatrix} a & b \\ d & c \end{bmatrix}$.

We proceed by assuming the row player chooses among his or her two choices with probability p; in other words, $[p \quad 1-p]$. Then, as before, we find the row player's average return when the column player uses column 1 or 2. If the column player chooses the first column with probability q, his or her average losses are given when the row player uses rows 1 and 2, that is, $[q \quad 1-q]$.

With no saddle point, (a – b) and (c – d) are either both positive or both negative, so p is strictly between 0 and 1. If $a \geq b$, then $b < c$; otherwise, b is a saddle point. Since $b < c$, we must have $c > d$, as otherwise c is a saddle point. Similarly, we can see that $d < a$ and $a > b$. This leads us to: if $a \geq b$, then $a > b < c > d < a$. Using an argument by symmetry, if $a <= b$, then $a < b > c < d > a$. And therefore:

If there is no saddle point, then either $a > b$, $b < c$, $c > d$ *and* $d < a$, *or* $a < b$, $b < c$, $c < d$ *and* $d > a$.

If the row player chooses the first row with probability p, the average return when the column player uses columns 1 and 2 is: $ap + d(1 - p) = bp + c(1 - p)$. Now solve this equation for p:

$$p = \frac{c-d}{(a-b)+(c-d)}$$

Since there is no saddle point, (a – b) and (c – d) are either both positive or both negative, so $0 < p < 1$. The row player's average return with this strategy is

$$v = ap + d(1-p) = \frac{ac - bd}{a-b+c-d}.$$

If the column player uses [q 1−q], his or her average losses are $aq + b(1 - q) = dq + c(1 - q)$. Thus

$$q = \frac{c-b}{a-b+c-d}.$$

Since there is no saddle point, $0 < q < 1$, so the column player's average loss is

$$aq + b(1-q) = \frac{ac - bd}{a-b+c-d} = v.$$

Since both players have the same optimal strategies, the game has a solution, which is the one given above.

Example 1

$$A = \begin{bmatrix} -4 & 6 \\ 6 & -8 \end{bmatrix}.$$

$$p = (-8-6)/(-4-6-8-6) = (-14)/(-24)$$
$$= 7/12 = 0.583 = q.$$

Example 2

$$A = \begin{bmatrix} 0 & -30 \\ 3 & 6 \end{bmatrix}.$$

$$p = (6-3)/(0+30+6-3) = (3)/(33) = 1/11 = 0.091$$

$$\text{But} \quad q = (6+30)/(0+30+6-3) = 12/11.$$

Since q must be between 0 and 1, we have made an error. In this case, we did not check first for a saddle point. Since the element $a_{21} = 3$, and it is the minimum in its row and the maximum in its column, it is a saddle point, so the correct solution is $p = 0$ and $q = 1$ as strategies.

13.6 DOMINATED STRATEGIES

It is often possible to reduce the size (i.e., the dimensions of the game matrix) by deleting rows or columns that are clearly losing strategies for the player who might choose the row or column.

Definition: We say one row of a game matrix dominates another if every value in the first row is greater than the corresponding value in the other row. Similarly, we say that a column of the matrix dominates another if every value in the first column is less than the corresponding value in the second column. More formally, we have the definition:

The ith row of a matrix $A = [a_{ij}]$ dominates the kth row if $a_{ij} \geq a_{kj}$ for all j. If we replace the greater than or equal to with the strictly greater than; in other words, $a_{ij} > a_{kj}$, we say that the ith row strictly dominates the kth row. Regarding the columns of the matrix, we say that the jth column of A dominates (strictly dominates) the kth column if $a_{ij} <= a_{ik}$ (respectively $a_{ij} < a_{ik}$) for all i.

Anything the row player can achieve by using a dominated row can be achieved at least as well by using the row that dominates. As such, a row player would never choose to play the dominated row, and a dominated row may be deleted from the matrix. Similarly, a column player can delete a dominated column from the matrix. In other words, the value of the game remains the same after dominated rows or columns are deleted.

It is possible that there may exist an optimal strategy that uses a dominated row or column; if that is the case, such a removal will also remove the use of such an optimal strategy. However, in the case of a removal of a strictly dominated row or column, the set of optimal strategies does not change.

Example: Consider the game matrix

$$A = \begin{bmatrix} 7 & 1 & 6 \\ 3 & 4 & 5 \\ 9 & 2 & 4 \end{bmatrix}.$$

We observe that the second column of A is strictly dominated by the third column, so we can effectively remove the third column from the game, leading to

$$A' = \begin{bmatrix} 7 & 1 \\ 3 & 4 \\ 9 & 2 \end{bmatrix}.$$

By a similar principle, the first row is strictly dominated by the third row, so the first row can be essentially eliminated. Thus, we are reduced to $A'' = \begin{bmatrix} 3 & 4 \\ 9 & 2 \end{bmatrix}$, and the resulting 2 × 2 matrix may be solved by the methods in the previous section. So, p, q, and v would be:

$$p = (c-d)/((a-b)+(c-d)) = (2-9)/((3-4)+(2-9))$$
$$= 7/8 = 0.875;$$
$$q = (c-b)/((a-b)+(c-d)) = (2-4)/((3-4)+(2-9))$$
$$= 1/4 = 0.25;$$
$$v = (ac-bd)/((a-b)+(c-d)) = (6-36)/(-1-7)$$
$$= 30/8 = 3.75.$$

Example: Colonel Blotto

One of the sources for examples in game theory arises from military strategy. One classic military strategy game is called Colonel Blotto.

Colonel Blotto must defend two mountain passes, and to do this, she has three divisions. She can successfully defend one of her mountain passes if she has a division that is pitted against an enemy unit of equal or smaller strength. The enemy, however, has only two divisions. Blotto will lose the battle if either pass is captured. Neither side has any advance information on how the opponent's divisions are allocated. The game, then, is described by the different possible alignments of both Blotto's divisions and the opponents.

We can describe the strategies for Blotto as having four possibilities, which we describe as (3,0), (2,1), (1,2), and (0,3). The first coordinate describes how many divisions Blotto assigns to the first mountain pass, and the second coordinate the number of divisions assigned to the second pass. Clearly, it wouldn't make any sense for Blotto to leave any of her divisions behind, so in each case Blotto is assigning all three divisions. The opponent, as indicated above, has only two divisions and thus has only three potential strategies: (2,0), (1,1), and (0,2).

From this information, we can develop the game matrix. Let's also assume that the payoff for the game is one unit, depending on who wins the battle. As an example, if Blotto assigns all of her divisions to the first pass, that is (3,0); and the opponent assigns one of his divisions to each pass, otherwise (1,1), the colonel loses the battle and the opponent wins; thus, the matrix entry is −1.

The battle is lost if either pass is captured. The payoff matrix is thus that shown in Table 13.3.

A row or column may also be removed if it is dominated by a probability combination of other rows or columns.

Just on preliminary observation of Colonel Blotto's game, her (3,0) and (0,3) strategies are never required, since the enemy only has two divisions, and Blotto can successfully defend any pass with only two divisions.

But we can reach the same conclusion through examination of the payoff matrix. Because the payoffs in the (3,0) row are all less than or equal to the numbers in the

TABLE 13.3
Colonel Blotto

		Enemy		
		(2,0)	**(1,1)**	**(0,2)**
Blotto	(3,0)	1	−1	−1
	(2,1)	1	1	−1
	(1,2)	−1	1	1
	(0,3)	−1	−1	1

TABLE 13.4
Colonel Blotto Game "Reduced"

		Enemy		
		(2,0)	**(1,1)**	**(0,2)**
Blotto	(2,1)	1	1	−1
	(1,2)	−1	1	1

TABLE 13.5
Colonel Blotto Game Second Reduction

		Enemy	
		(2,0)	**(0,2)**
Blotto	(2,1)	1	−1
	(1,2)	−1	1

same column in the second row, and similarly for the (0,3) row, by the principle of domination, we can reduce the game to Table 13.4.

Now, examining the game matrix from the column perspective, that is, the enemy's perspective, the strategy (1,1) is also dominated. Thus, we can also remove that column from the game (Table 13.5).

Considering the remaining 2 × 2 matrix, we see we have essentially reduced the Colonel Blotto game to the game of matching coin flips, and from our previous analysis, we realize that the players should each randomize their choice of strategy, leading to a 50–50 conclusion.

13.7 GRAPHICAL SOLUTIONS: 2 × n AND m × 2 GAMES

In a special case where there are either only two strategies for the row player (called a 2 × n game) or only two strategies for the column player (called an m × 2 game), there is an approach that can be used to solve the game using a graphical representation. With the following 2 × n game, we can proceed as follows. Clearly, the row player chooses the first row with probability p; then, that player's strategy for the choice of second row must be 1 – p. With the example above, we can calculate the average payoff for the column player with each of the n strategies available to him or her. Those payoffs are described as follows:

We can plot the potential values for the game outcome by graphing the straight line for the values of p between 0 and 1.

(A student familiar with the subject of linear programming will recognize the solution method as similar to the simplex method.)

Consider the following example for a two-row, four-column game (in other words, a 2 × n game):

p	6	8	2	9
1–p	8	4	10	2

If the column player chooses the first column, the average payoff for the row player is 6p + 8(1 – p). By the same token, choices of the second, third, and fourth columns would lead to average payoffs of: (column 2) 8p + 4(1 – p); (column 3) 2p + 10(1 – p); (column 4) 9p + 2(1 – p). Next, graph all four of these linear functions for p going from 0 to 1 (Figure 13.1).

For any fixed value of p, the row player is sure that his or her average winnings are at least the minimum of these four functions evaluated for the chosen value of p. Thus, in this range for p, we want to find the p that achieves the maximum of this "lower envelope." In geometry or linear programming, this lower envelope is also referred to as the convex hull. In this example, we can see that maximal value for the lower envelope occurs when p = 8/15. Because the only columns that intersect at the critical point are columns 3 and 4, we can also conclude that we can reduce the game to the second and third columns, therefore once again reducing us to the 2 × 2 game for which we already have a methodology for a solution.

We can verify our graphical solution by using the algebraic approach for the solution of the reduced 2 × 2 game.

We can follow a similar approach for an m × 2 game. Suppose this time the game is

q	1–q
7	2
3	6
5	5

Once again, we plot the three line segments for q between 0 and 1, and this time take the "upper envelope" and determine the values of q at the critical points (Figure 13.2).

In this case, we will find that there are two strategies, one for the value q = 1/3 and one for the value q = 3/5, in each case leading to a payoff of 5.

Thus, this methodology can be used to solve any 2 × n or m × 2 game.

In theory, we can extend this graphical method to games that are 3 × n or m × 3. However, for the example in which the game has three rows instead of two, the

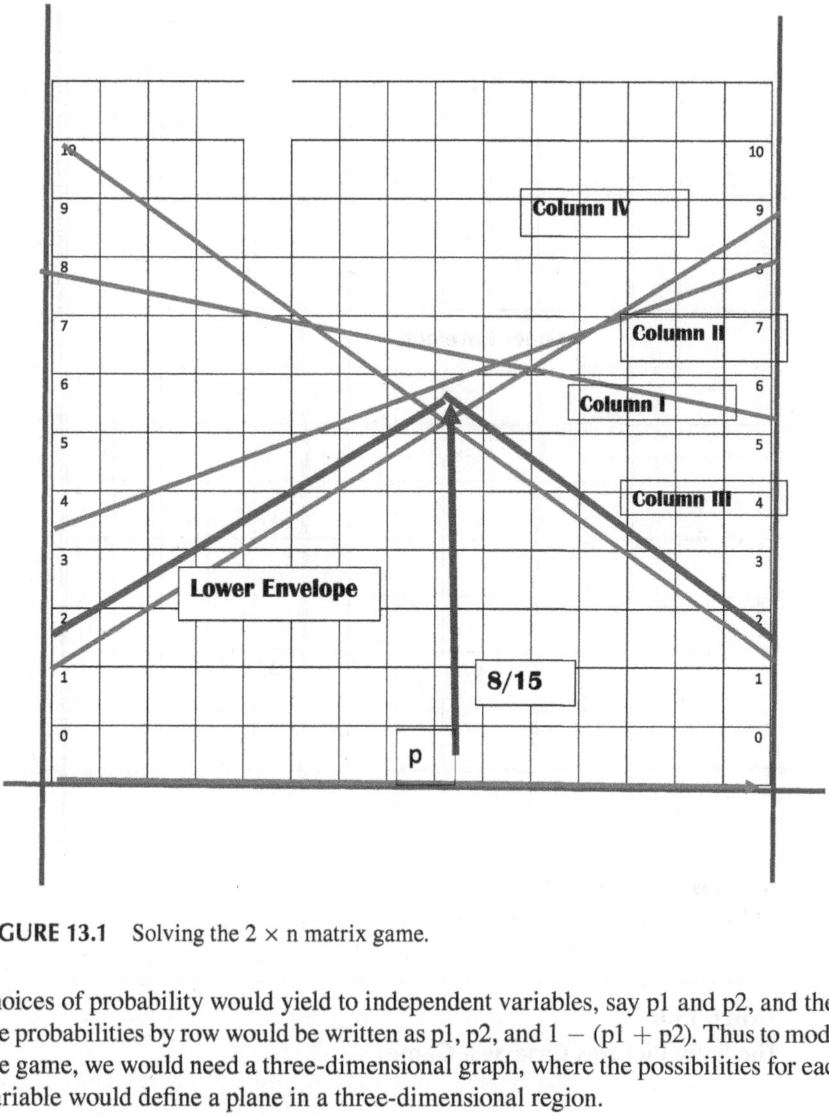

FIGURE 13.1 Solving the 2 × n matrix game.

choices of probability would yield to independent variables, say p1 and p2, and then the probabilities by row would be written as p1, p2, and 1 − (p1 + p2). Thus to model the game, we would need a three-dimensional graph, where the possibilities for each variable would define a plane in a three-dimensional region.

13.8 USING GAME THEORY TO CHOOSE A STRATEGY IN THE SONY/NORTH KOREA CASE

Recall the case of the Sony Pictures hack from Chapter 1 (Sony, 2018). We consider the potential attackers we discussed as all being coordinated by some criminal mastermind, whom we will abbreviate CM. CM will play the game by choosing one of the potential attackers we considered the real attacker. How CM makes this choice will be his or her strategy. CM will be the row player.

The cybersecurity expert (you) will be abbreviated CE and will try to determine the motivation for the attack. CE will be listed across the columns (Table 13.6).

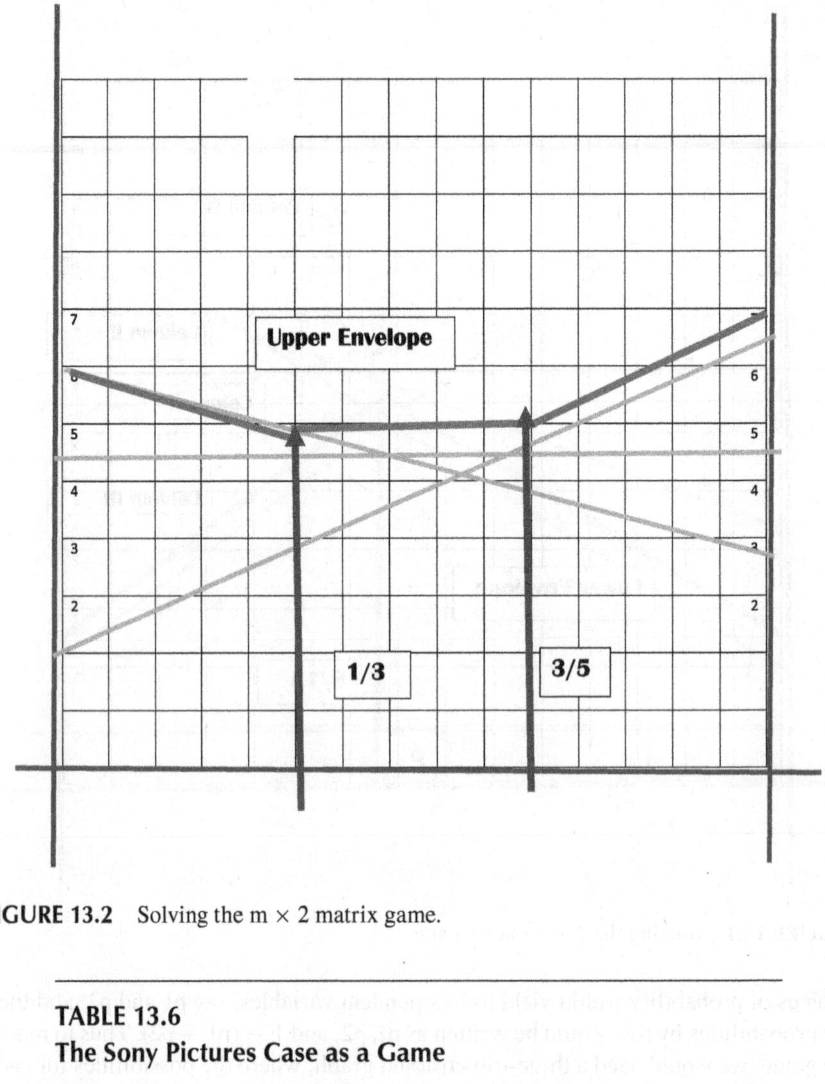

FIGURE 13.2 Solving the m × 2 matrix game.

TABLE 13.6
The Sony Pictures Case as a Game

	CE			
CM	**Politics**	**Warfare**	**Reputation**	**Money**
North Korea				
Guardians of Peace				
WikiLeaks				
Industrial Competitor				

We assume that we know the payoffs for each possibility, as described in the matrix below. The game will be to find the optimal strategy (best probability).

I will treat CM as the row player and CE as the column player. The payoff matrix looks like Table 13.7.

TABLE 13.7
The Sony Pictures Case as a Game for Solution

	CE			
CM	**Politics**	**Warfare**	**Reputation**	**Money**
North Korea	1	−1	−1	−1
Guardians of Peace	−1	−1	1	−1
WikiLeaks	−1	1	−1	−1
Industrial Competitor	−1	−1	−1	1

The rules of play are that if both CM and CE choose (Attacker, Motivation) = (North Korea, Politics), then CM wins (one unit of something).

CM also wins if neither chooses (North Korea, Politics) and CE guesses differently from CM. CE wins otherwise.

Using pure strategies, neither player can guarantee a win. The pure strategy security level for each player is therefore −1. Figuring out the mixed security strategy level "by hand" for a game with four strategies for each player is tedious and hard. In this example, three of the strategies are symmetric. This suggests a simplification.

Assume that CM always plays Guardians of Peace, WikiLeaks, Industrial Competitor, with the same probability. (This probability can be any number between 0 and 1/3; if the probability is zero, then CM always plays North Korea; if the probability is 1/3, then CM never plays North Korea, and plays each of the remaining Attackers with the same probability. You can view the game as heading to pure strategies for CM (either he plays Sony or he doesn't). The payoff matrix becomes:

	Politics	Warfare	Reputation	Money
North Korea	1	−1	−1	−1
Not North Korea	−1	1/3	1/3	1/3

If you impose the same symmetry condition on CE, the game reduces to Table 13.8.

It is now possible to find the mixed-strategy equilibrium for this 2 × 2 (i.e., two strategies for each player) game. CM's strategy will equalize the payoff she gets from either strategy choice of CE. That is, the probability of (North Korea, Politics), call it a, will satisfy:

$$a + (1 - a)(-1) = a(-1) + (1 - a)(1/3)$$

TABLE 13.8
The Sony Pictures Game Solution

	Politics	Not Politics
North Korea	1	−1
Not North Korea	−1	1/3

The solution to this equation is a = 2/5. When CM mixes between North Korea and Not North Korea with probabilities 2/5 and 3/5, he gets a payoff of –1/5 whatever CE does. Similarly, you can solve for CE's strategy by equalizing CM's payoff. If b is the probability that CE plays (North Korea, Politics), then b should satisfy:

$$b + (1 - b)(-1) = b(-1) + (1 - b)(1/3)$$

or

$$b = 2/5.$$

Now you can go back and check that the symmetry assumption that I imposed is really appropriate. Both players are playing (North Korea, Politics) with probability 2/5 and other strategies with total probability 3/5. That means that they play each of the (Non-North Korea, Politics) strategies with probability 1/3. Under this, CM expects to earn –1/5 from each of his original pure strategies, and CE can hold CM to this amount by playing (North Korea, Politics) with probability 2/5 and the other three strategies with probability of 1/5 each.

13.9 TRANSFORMING A PROFILING MATRIX TO A GAME THEORY PROBLEM

From a different perspective, we will use the Sony example again to show a method of transforming a profiling matrix problem to a game theory problem. In Chapter 5, we describe the methodology for profiling suspects in a cyberattack using a technique called profiling matrices. Using the Sony example that analyzed the 16 suspects we developed that were identified, along with the 12 motivations identified, we will construct a game theory problem.

The techniques of two-person zero-sum game theory can be used to solve the profiling matrix by converting it to a game theory problem.

Just for the sake of this example, will use the data developed for the Sony hack that was developed by our Howard University students.

We first reinterpret the profiling matrix as a game, in this case with the row player having 16 strategies (suspects) and the column player 12 strategies (motivations).

We will also assume that we have made some progress in reducing that case and that we are left with two potential suspects, GOP and Russia, and that we have reduced the number of motivations to seven, which we label in short form Politics (P), War (W), Reputation (R), Conspiracy (C), Fame (F), Vendetta (V), and Information (I).

Using the specific data from Chapter 5, we now have the following reduction to a 2 × 7 game. (The figures are percentages.)

		P	W	R	C	F	V	I
GOP	p	12.9	4.3	41.1	16.4	23.2	9.6	12.4
Russia	1 – p	23.5	36.7	2.7	31.0	1.9	0.7	6.6

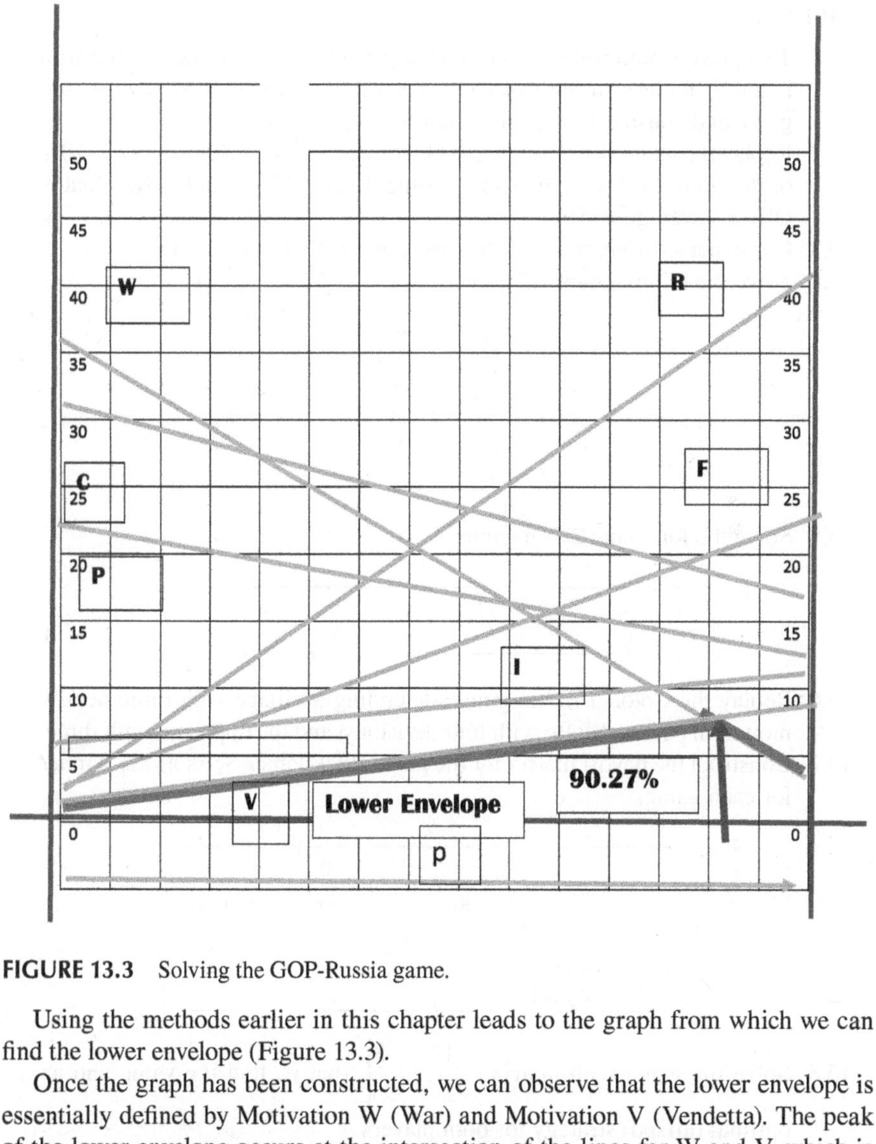

FIGURE 13.3 Solving the GOP-Russia game.

Using the methods earlier in this chapter leads to the graph from which we can find the lower envelope (Figure 13.3).

Once the graph has been constructed, we can observe that the lower envelope is essentially defined by Motivation W (War) and Motivation V (Vendetta). The peak of the lower envelope occurs at the intersection of the lines for W and V, which is approximately at the point p = 90.78. Therefore, we can conclude that there is a 90.78% chance that the likely suspect is Russia with the Motivation War.

REFERENCES

Morris, P. 1994. *Introduction to Game Theory.* New York: Springer, July 28.
Sony Pictures Digital Productions Inc. 2018. http://www.sonypictures.com/
Von Neumann J. and Morgenstern, O. 1944. *Theory of Games and Economic Behavior.* Princeton, NJ: Princeton University Press.

PROBLEMS

13.1 Two players both roll the dice. The larger roll wins, and the amount won is the difference in the two rolls. Treat this as a two-person, zero-sum game and construct the payoff matrix.

13.2 Replay the game with Evelyn and Oliver where Oliver chooses "one" 70% of the time and "two" 30% of the time. Evelyn plays randomly. What is Oliver's average win?

13.3 Construct a different 4 × 4 strategy game with a saddle point.

13.4 Solve the following m × 2 game:

3	6
1	4
7	2
3	9
4	7

13.5 Solve the following 2 × n game.

4	8	3	1	6	8
6	2	7	8	2	7

13.6 Replay the Colonel Blotto game, defending a village with three nearby mountain passes, Blotto with four divisions, and the opponent with three.

13.7 Construct the payoff matrix for the game Rock-Paper-Scissors ($1 winner for each game).

		II		
		Rock	Paper	Scissors
	Rock			
I	Paper			
	Scissors			

13.8 Solve the game with matrix $\begin{bmatrix} -1 & -3 \\ -2 & 2 \end{bmatrix}$; that is, find the value and an optimal (mixed) strategy for both players.

13.9 Reduce by dominance to a 3 × 2 matrix game and solve: $\begin{bmatrix} 0 & 8 & 5 \\ 8 & 4 & 6 \\ 12 & -4 & 3 \end{bmatrix}$.

14 Ethical Hacking

A hacker is a person with the technical skill and knowledge to break into computer systems, access files and other information, modify information that may be in the computer system, use skills involving network technology to move from one system to another, and implant software that may have deleterious effects on the host system.

An ethical hacker is a person with the technical skill and knowledge to break into computer systems, access files and other information, modify information that may be in the computer system, use skills involving network technology to move from one system to another, and implant software that may have deleterious effects on the host system.

What is the difference in the abilities of the hacker and the ethical hacker? Well, as you are easily able to observe, there is essentially no difference.

The International Council of Electronic Commerce Consultants (EC-Council) is a professional certification body that maintains a process of certification as a Certified Ethical Hacker (CEH, 2018). A person gaining this certification will have the right to refer to himself or herself by this term; however, many people who are employed in protecting cyberenvironments do so without the aid of this official designation.

Given the fact that there is essentially no difference in the technical skill set of a hacker or an ethical hacker, one might wonder what difference in fact can there be. It is also the case that other terminology has become widespread: the hackers are often called "black hats" and the ethical hackers "white hats." However, to further confuse the issue, there are competitions wherein the participants are assigned to the white hat or black hat team, but in midcompetition they may change hats and roles.

This terminology is somewhat unusual. In other areas of human activity where we consider behavior either legal or illegal, ethical or unethical, it stretches the imagination to consider, for example: bank robbery or ethical bank robber, murderer or ethical murderer. If Robin Hood were indeed an actual person, he would probably have liked to be considered an ethical thief.

In the realm of cybersecurity, it seems that we have evolved to the point where the main distinction between the hacker and the ethical hacker has nothing to do whatsoever with skill or ability but rather the values of the individual falling into one or the other category.

A distinguished computer scientist named Ymir Vigfusson, originally from Iceland and more recently a professor at Emory University in Atlanta, offers courses in ethical hacking and has described his philosophy extremely well in a recent Ted Talk, "Why I Teach People How To Hack" (https://youtu.be/KwJyKmCbOws) (Vigfusson, 2015).

Prof. Vigfusson uses the term "moral compass" to describe what guides him as a professor in teaching ethical hacking to his students and also how he operates in his own practice.

Thus, it seems that the challenge in the cybersecurity profession is to find a way of identifying how an individual who can develop the requisite technical skills can rely on his or her own moral compass.

An analogy that might be useful in this regard arises from an experience over many years of the first author. There was for many years a very important international development program of the United States Agency for International Development, originally called AFGRAD and later ATLAS. The purpose of the program was to provide masters and doctoral fellowships for sub-Saharan African students to come to do their advanced degrees at United States universities. A committee of graduate school deans was responsible for evaluating and proposing the candidates for selection in the various African countries. The first author served on this committee for about 10 years.

The criteria for selection in this program was of course first that the candidate be capable of successfully completing the degree at the appropriate United States university.

But, as it developed, equal in importance in our judgment (since this was a program for international development) was that we attempted to evaluate the candidates in terms of our perception of their commitment to return to their home country after their degree, and not use the entrée to the United States to give them a potentially more lucrative environment. The fact that we took this criterion so seriously is evident in the eventual results: of almost 4000 students awarded these scholarships, over 90% completed their graduate degrees in the United States and returned to their home countries in sub-Saharan Africa to contribute to their eventual development (USAID, 1995), what we might describe as our assessment of their moral compass.

This seems to be a pattern that could be used in evaluating candidates for developing their skills in cybersecurity. How can we judge their "moral compass" to predict whether they will use their skills in cybersecurity in defensive roles, in other words, in deterring cyberattacks, or whether they will be tempted to be an offensive attacker by gaining personal wealth, success, reputation, or other attributes that might be as tempting as the ATLAS students' temptation to live in an economically wealthier society might have been.

Developing measures to try to predict these behaviors is a clear challenge for those persons who are not only knowledgeable about cybersecurity but also about psychology and the behavioral sciences.

14.1 PROGRAMS TO ENCOURAGE THE DEVELOPMENT OF ETHICAL HACKERS

Very recently, greater attention has been drawn to initiatives that attempt to encourage the development of ethical hackers.

For example, on December 8, 2015, Donna Lu wrote in *The Atlantic*:

> The cybersecurity expert Chris Rock (not the Saturday Night Live comedian) is an Australian information-security researcher who has demonstrated how to manipulate online death-certification systems in order to declare a living person legally dead.
>
> Brock began researching these hacks last year, after a Melbourne hospital mistakenly issued 200 death certificates instead of discharge notices for living patients. He also uncovered similar vulnerabilities in online birth registration systems. The ability to

create both birth and death certificates, Rock told a packed session at DEF CON, meant that hackers could fabricate new legal identities, which could in turn engender new types of money-laundering and insurance fraud schemes.

In the hacking world, Rock is known as a "white hat": an ethical hacker who exposes vulnerabilities in computer systems to improve cybersecurity, rather than compromise it. In recent years white-hat hacking has become increasingly lucrative, as companies have turned to professionals like Rock to protect them from the growing threat of cyber crime. But to combat the sophistication of more malevolent hackers, the ethical-hacking industry still has a long way to go.

Subsequently, on August 2, 2017, Kevin Roose in the *New York Times* wrote:

If there is a single lesson Americans have learned from the events of the past year, it might be this: hackers are dangerous people. They interfere in our elections, bring giant corporations to their knees, and steal passwords and credit card numbers by the truckload. They ignore boundaries. They delight in creating chaos.

But what if that's the wrong narrative? What if we're ignoring a different group of hackers who aren't lawless renegades, who are in fact patriotic, public-spirited Americans who want to use their technical skills to protect our country from cyber-attacks, but are being held back by outdated rules and overly protective institutions?

In other words: What if the problem we face is not too many bad hackers, but too few good ones?

The topic of ethical hacking was on everyone's mind at DEF CON, the hacker convention last week in Las Vegas. It's the security community's annual gathering, where thousands of hackers gathered to show their latest exploits, discuss new security research and swap cyber war stories. Many of the hackers I spoke to were greatly concerned about Russia's wide-ranging interference in last year's election. They wanted to know: How can we stop attacks like these in the future?

The problem, they told me, is that government doesn't make it easy for well-meaning hackers to pitch in on defense. Laws like the Computer Fraud and Abuse Act make poking around inside many government systems, even for innocent research purposes, a criminal offense. More than 209,000 cybersecurity jobs in the United States currently sit unfilled, according to a 2015 analysis of labor data by Peninsula Press, and the former head of the National Security Agency said last year that the agency's cybersecurity experts "are increasingly leaving in large numbers" for jobs in the private sector.

And most recently, on November 24, 2017, Anna Wiener wrote in the *New Yorker*:

"Whenever I teach a security class, it happens that there is something going on in the news cycle that ties into it," Doug Tygar, a computer-science professor at the University of California, Berkeley, told me recently. Pedagogically speaking, this has been an especially fruitful year. So far in 2017, the Identity Theft Resource Center, an American nonprofit, has tallied more than eleven hundred data breaches, the highest number since 2005. The organization's running list of victims includes health-care providers, fast-food franchises, multinational banks, public high schools and private colleges, a family-run chocolatier, an e-cigarette distributor, and the U.S. Air Force. In all, at least a hundred and seventy-one million records have been compromised. Nearly eighty-five per cent of those can be traced to a single catastrophic breach at the credit-reporting agency Equifax. That hack was reported in early September—just as Tygar and his students were settling into the third week of a new course called "Cyberwar."

The purpose of the course, according to Tygar's faculty Web page, is to teach Berkeley's budding computer scientists to "forensically examine real cyberwar attacks" with an eye toward preventing them. Occasionally, this might mean mounting attacks of their own. Penal codes around the U.S. are not especially lenient when it comes to cybercrime; in some states, certain computer crimes are considered Class C felonies, on par with arson and kidnapping. So, for the hands-on portion of their studies, Tygar's students rely on HackerOne, a sort of marketplace-cum-social-network devoted to "ethical hacking." Companies, organizations, and government agencies use the site to solicit help identifying vulnerabilities in their products—or, as Tygar put it, "subject themselves to the indignity of having undergraduate students try to hack them." In exchange for information about what they're doing wrong, many of these clients offer monetary rewards, known as bug bounties. Since 2012, when HackerOne was launched, its hundred thousand or so testers have earned a total of twenty-two million dollars, a figure that the platform's Dutch-born founders, Jobert Abma and Michiel Prins, hope to quintuple by 2020. For Tygar's students, there is an added incentive: every bug they catch through HackerOne also gets them points toward their final grades.

REFERENCES

Certified Ethical Hacker (CEH). 2018. International Council of Electronic Commerce Consultants.

Lu, D. 2015. When ethical hacking can't compete. *The Atlantic*, December 8.

Roose, K. 2017. A solution to hackers? More hackers. *New York Times*, August 2.

United States Agency for International Development (USAID) 1995. *AFGRAD III Evaluation Report: Capturing the Results of 30 Years of AFGRAD Training.* USAID Project No. 698-0455, December.

Vigfusson, Y. 2015. *Why I Teach People How to Hack*. Ted Talk, March 24.

Wiener, A. 2017. At Berkeley, a new generation of 'ethical hackers' learns to wage cyberwar. *New Yorker*, November 24.

PROBLEMS

14.1 Find the CEH requirements.

14.2 Find the CEH test questions.

14.3 Watch the YouTube and Ted Talk by Ymir Vigfusson. What part of discovering his hack brought him joy, and then respond to his "moral compass"?

14.4 Find the origin of the terms "white hat" and "black hat."

14.5 Find the number of convictions under the Computer Fraud and Abuse Act.

14.6 Comment on Roose's *New York Times* article: "What if the problem we face is not too many bad hackers, but too few good ones?"

14.7 Update the information from the Anna Wiener article regarding the number of data breaches from the Identity Theft Resource Center.

15 The Psychology of Gender

The use of a private email server by Hillary Rodham Clinton while she was secretary of state during the presidential administration of Barack Obama has sparked a reoccurring debate about the use of private email servers by U.S. government officials. In 2016, Hillary Rodham Clinton was the first woman to win the Democratic Party's nomination for the president of the United States. As is the case in many political situations, there are some angles of the issue that can be more fully explored to understand the nature of the issue. An interesting question to explore the landscape of gender psychology and behavioral cybersecurity is as follows: How might different common conceptualizations of gender along with corresponding approaches to understanding the psychology of gender explain the behavior of Secretary of State Hillary Clinton and what caused her use of a private email server? The purpose of this chapter is to describe theoretical conceptualizations of the psychology of gender to advance the argument that these conceptualizations can be used to better understand, protect, and defend information and systems in cybersecurity situations and cases. These conceptualizations can also be used to advance understanding of the participation of women in computer science and other computing fields. This chapter is organized with a brief snapshot of the background and historical context of approaches to understanding gender in psychology, followed by a more detailed description of these dominant approaches to gender psychology, and concludes with a nature vs. nurture framework for making sense of the different approaches.

15.1 BACKGROUND AND HISTORICAL CONTEXT: GENDER AND PSYCHOLOGY

Since the founding of the field of psychology, there has been interest in understanding the behavior of women and men. Many popular historical and contemporary psychological theories describe, explain, or predict gendered origins and/or patterns in the behavior of men and women. These theories range from Freud's theory of unconscious drives and the structure of personality that became popular in the early twentieth century to Buss's (1995) more contemporary theory of human evolution focused on mate selection that emerged in the 1990s.

Arising in the 1960s, feminist psychologists had a major impact on challenging the academic and clinical formulations of the nature of the female that were popular prior to this time (Shields and Dicicco, 2011). Thus, they began to shift approaches to understanding the psychology of gender through a move from questions and theoretical frameworks centered on sex-related differences in psychological processes and outcomes to a more social-contextual view of gender (e.g. Deaux and Major, 1987; Fine and Gordon, 1991; Spence and Helmreich, 1981; Marecek, 2001; Spence et al., 1975).

Shields and Dicicco (2011) described the extension of this socio-contextual view of gender during the 1970s by both feminist psychologists and other researchers who introduced the idea that there was a need for explicit differentiation "between sex as categorization on the basis of anatomy and physiology, and gender as a culturally defined set of meanings attached to sex and sex difference" (p. 493). Theorizing and research on prenatal and postnatal gender development had a profound influence on this shift in differentiation between sex and gender psychological conceptualizations (Money and Erhardt, 1972; Spence and Helmreich, 1978; Unger, 1979). Following these theoretical and empirical developments in psychology, there was a wave of seminal scholarship on the socio-contextual views of the psychology of gender (Deaux and Major, 1987; Eagly, 1987, 1994; Eagly and Wood, 1999; Eccles, 1987; West and Zimmerman, 1987) that has persisted today. In addition, more current psychological research on gender continues to debate if more complex formulations of similarities or differences would be more theoretically and practically robust for advancing understanding of the psychology of gender (Fine and Gordon, 1991; Shields and Bhatia, 2009).

15.2 THE CONCEPTUALIZATION AND ANALYSIS OF GENDER

Within behavioral science, there is theoretical and methodological variation in how gender is conceptualized and studied. However, there are some dominant patterns that have been identified by psychologists in various theoretical and empirical review articles. For example, Stewart and McDermott (2004) summarized the gender psychology body of scholarship by identifying the following theoretical orientations: (1) gender as sex differences on outcomes, (2) gender as a role and gendered socialization, (3) gendered power relations, and (4) intersections of gender identity with other social and socio-structural identities (e.g., race, ethnicity, sexual orientation).

15.2.1 Gender-As-Trait: The Sex Differences Approach

Within this approach to gender psychology, the central question that psychologists pursue is: How and why do average differences in attitudes, ability, personality traits, and other behavioral tendencies appear? This approach assumes that differences arise from preexisting "essential" differences between males and females. Shields and Dicicco (2011) described that the core idea of this approach is that differences between male and females are "natural, deep-seated, and of profound personal and social consequences. This proposal easily built upon Anglo-American belief in 'natural' gender differences as differentiating 'advanced' races from more primitive" (p. 491). In essence, this approach views differences between males and females as being due to genes and hormones (Kitzinger, 1994).

15.2.2 Gender in Social Context: The Within-Gender Variability Approach

In this approach to gender psychology, the central question that psychologists pursue is: Within highly gendered psychological phenomena, what are the sources of within-gender variation? Highly gendered psychological phenomena refer to those phenomena defined by average differences found by researchers between men and women (i.e., attitudes, ability, motivation, personality traits, and other behavioral tendencies). Eccles et al. 1983, Eccles (2007, 2009, 2011), for example, in her theoretical and empirical work on expectancy-value theory, demonstrated how the social environment shapes individuals' expectation of success, ideas about the importance of a task, and perception of available options, as well as academic and career choices in science and mathematics disciplines/fields.

15.2.3 Gender Linked to Power Relations Approach

In this approach to gender psychology, the central question that psychologists pursue is: How does gender structure social institutions' practices, norms, and policies within which men and women operate? With this approach, there is a recognition that gender beliefs and behaviors are ideologies that are embedded in social-structural systems (Shields and Dicicco, 2011). Stewart and McDermott (2004) described this approach by explaining that "gender" operates not merely at the level of sex differences, or as the result of social interactions in which beliefs about gender are expressed in actions that actually create confirming evidence for those beliefs, but also in the social structures that define power relationships throughout culture (p. 521; examples of such scholarship are Fiske and Stevens, 1993; Goodwin and Fisk, 2001; Stewart, 1998). The kinds of topics that psychologists have explored within this approach are leadership, marital relationships, decision-making (i.e., choice) and conflict, and task performance. There are various configurations of social relationships in which these behaviors take place, including dyads, organizational hierarchies, and broader cultural political structures (Deaux and Major, 1987).

15.2.4 Gender as Intersectional: The Identity Role, Social Identity, and Social Structural Approach

Within this approach to gender psychology, the central question that psychologists pursue is: How do gender roles, social identities, and social structural dynamics operate individually and interactionally to shape psychological processes and outcomes of men and women? As Stewart and McDermott (2004) explained in their seminal review of gender psychology, this approach can adopt distinctive theoretical orientations grounded in individual identity theory rooted primarily in personality psychology; social identity theory anchored largely within social identity theory; racial identity theory cutting across personality, social, and developmental psychology; and intersectionality theory that encompasses boundaryless subareas of psychology. The kinds of topics that psychologists have explored within this approach are ego identity development, gender role identity (Eccles, 2007), social identity (Gurin, 1985; Gurin and Markus, 1989), racial and ethnic identity (e.g., Sellers et al., 1998), and intersectionality (e.g., Cole, 2009; Crenshaw,1994; Hurtado and Sinha, 2008; Ireland, Freeman, Winston-Proctor, DeLaine, Lowe, and Woodson, 2018; Mack, Rankins and Winston, 2011; Shields, 2008). One example of gender role theory was the research of psychologists who explored gender socialization in terms of the experience of having one's behavior, beliefs, and attitudes shaped by culturally defined, gender-specific roles (Shields and Dicicco, 2011). For example, increasingly, psychologists are adopting feminist theories and critical race theories to develop theoretical, methodological, and practical formulations of intersectionality, which refers to the simultaneous meaning and consequences of multiple categories of identity, difference, and disadvantage, particularly related to the intersections of race, gender, and social class (Cole, 2009).

15.3 THE NATURE VS. NURTURE DEBATE IN GENDER PSYCHOLOGY

One common way to classify these dominant approaches to the psychology of gender is as nature, nature, or a combination of the two. This classification represents a classic debate across the history of the field of psychology. The fundamental question that undergirds this debate and thus informs the various approaches to gender psychology is as follows: Is nature or nature responsible for differences and similarities found in the beliefs, attitudes, abilities, motives, personality traits, and other behavioral tendencies of men and women? Key constructs examined related to nurture are those grounded in sociocultural influences, while those anchored in nature are biological structural processes. Eagly and Wood (2013) pointed out that it is common for researchers to focus on one cause to the exclusion of the other or to treat them as competing explanations.

15.4 CONCLUSION

Gender has had and has diverse meanings, but nature vs. nurture debates persist, though many psychologists and other researchers who study the psychology of gender are advocating for a more integrative approach. Especially promising are new areas of neuropsychology that are opening up new theories and directions for the study of gender in the field of behavioral science.

REFERENCES

Buss, D. M. 1995. Psychological sex difference: Origins through sexual selection. *American Psychologist*, 50(3), 164–168.

Cole, E. R. 2009. Intersectionality and research in psychology. *American Psychologist*, 64(3), 170–180.

Crenshaw, K. W. 1994. Mapping the margins: Intersectionality, identity politics, and violence against women of color. In M. A. Fineman and R. Mykitiuk (Eds.), *The Public Nature of Private Violence* (pp. 93–118). New York: Routledge.

Deaux, K. and Major, B. 1987. Putting gender into context: An interactive model of gender-related behavior. *Psychological Review*, 94(3), 369–389.

Eagly, A. H. 1987. *Sex Differences in Social Behavior: A Social-Role Interpretation*. Hillsdale, NJ: Erlbaum.

Eagly, A. H. 1994. On comparing women and men. *Feminism and Psychology*, 4, 513–522.

Eagly, A. and Wood, W. 1999. The origins of sex differences in human behavior: Evolved dispositions versus social roles. *American Psychologist*, 54, 408–423.

Eagly, A. H. and Wood, W. 2013. The nature–nurture debates: 25 Years of challenges in understanding the psychology of gender. *Perspectives on Psychological Science*, 8, 340–357.

Eccles, J. S. 2011. Understanding women's achievement choices: Looking back and looking forward. *Psychology of Women Quarterly*, 35(3), 510–516.

Eccles J. S., Adler, T. F., Futterman, R., Goff, S. B., Kaczala, C. M., Meece, J. L., and Midgley, C. 1983. Expectancies, values and academic behaviors. In J. Spence (Ed.), *Achievement and Achievement Motivation* (pp. 75–146). San Francisco: W.H. Freeman and Co.

Fine, M. and Gordon, S. M. 1991. Effacing the center and the margins: Life at the intersection of psychology and feminism. *Feminism and Psychology*, 1(1), 19–27.

Fiske, S. T. and Stevens, L. E. 1993. What's so special about sex? Gender stereotyping and discrimination. In S. Oskamp and M. Costanzo (Eds.), *Gender Issues in Contemporary Society: Applied Social Psychology Annual* (pp. 173–196). Newbury Park, CA: Sage.

Goodwin, S. A. and Fiske, S. T. 2001. Power and gender: The double-edged sword of ambivalence. In R. K. Unger (Ed.), *Handbook of the Psychology of Women and Gender* (pp. 358–366). New York: Wiley.

Gurin, P. 1985. Women's gender consciousness. *Public Opinion Quarterly*, 49(2), 143–163.

Gurin, P. and Markus, H. 1989. The cognitive consequences of gender identity. In S. Skevington and D. Baker (Eds.), *The Social Identity of Women* (pp. 152–172). Sage Publications.

Hurtado, A. and Sinha, M. 2008. More than men: Latino feminist masculinities and intersectionality. *Sex Roles*, 59 (5–6), 337–349.

Ireland, D., Freeman, K. E., Winston-Proctor, C. E., DeLaine, K. D., Lowe, S. M., and Woodson, K. M. 2018. (Un)Hidden figures: A synthesis of research addressing the intersectional experiences of Black women and girls in STEM education. *Review of Research in Education*, 42(1), 226–254.

Kitzinger, C. 1994. Should psychologists study sex differences? *Feminism and Psychology*, 4(4), 501–506.

Mack, K. M., Rankins, C. M., and Winston, C. E. 2011. Black women faculty at historically Black colleges and universities: Perspectives for a national imperative. In H. T. Frierson and W. F. Tate (Eds.), *Beyond Stock Stories and Folk Tales: African Americans' Paths to STEM Fields (Diversity in Higher Education, Volume 11)* (pp. 149–164). Bingley, UK: Emerald Group Publishing Limited.

Marecek, J. 2001. After the facts: Psychology and the study of gender. *Canadian Psychology/Psychologie Canadienne*, 42(4), 254–267.

Money, J. and Ehrhardt, A. 1972. Man and Woman, Boy and Girl: The Differentiation and Dimorphism of Gender Identity from Conception to Maturity. Retrieved from http://proxyhu.wrlc.org/login?url=http://search.ebscohost.com/login.aspx?direct=trueanddb=sihandAN=SN084916andsite=ehost-live

Sellers, R. M., Smith, M. A., Shelton, J.N., Rowley, S. A. J., and Chavous, T. M. 1998. Multidimensional model of racial identity: A reconceptualization of African American racial identity. *Personality and Social Psychological Review*, 2, 18–39.

Shields, S. A. 2008. Gender: An intersectionality perspective. *Sex Roles*, 59, 301–311.

Shields, S. A. and Bhatia, S. 2009. Darwin and race, gender, and culture. *American Psychologist*, 64, 111–119.

Shields, S. A. and Dicicco, E. C. 2011. The social psychology of sex and gender: From gender differences to doing gender. *Psychology of Women Quarterly*, 35(3), 491–499.

Spence, J. T. and Helmreich, R. L. 1978, *Masculinity and Femininity: Their Psychological Dimensions, Correlates, and Antecedents*. Austin, TX: University of Texas Press.

Spence, J. T. and Helmreich, R . L . 1981. Androgyny versus gender schema: A comment on Bern's gender schema theory. *Psychological Review*, 88, 365–368.

Spence, J. T., Helmreich, R. and Stapp, J. 1975. Ratings of self and peers on sex-role attributes and their relations to self-esteem and conceptions of masculinity and femininity. *Journal of Personality and Social Psychology*, 32, 29–39.

Stewart, A. J. 1998. Doing personality research: How can feminist theories help? In B. M. Clinchy and J. K. Norem (Eds.), *Gender and Psychology Reader* (pp. 54–68). New York: New York University Press.

Stewart, A. and McDermott, C. 2004. *Gender in psychology. Annual Review of Psychology*, 55, 519–544.

Unger, R. K. 1979. Toward a redefinition of sex and gender, *American Psychologist* 34(11), 1085–1094.

West, C. and Zimmerman, D. G. 1987. Doing gender. *Gender and Society*, 1, 125–151.

PROBLEMS

These problems are designed to make you think about the essential behavioral science concepts that have been discussed in this chapter. These problems could be used in a number of ways, including as individual thought exercises, group discussion questions, and/or to simulate interest in new ways of thinking about gender psychology and behavioral cybersecurity.

15.1 How might different common conceptualizations of gender along with corresponding approaches to understanding the psychology of gender explain the behavior of Secretary of State Hillary Clinton and what caused her use of a private email server?

15.2 How might the approaches within behavioral science to gender psychology be used to advance understanding of the participation of women in computer science and other computing fields?

15.3 What are the different ways to think about how the approaches to gender psychology described in this chapter could be applied to analyze the Eliza program dialogue and the purpose described in Chapter 16, "Turing Tests"?

16 Turing Tests

A theme that we have continued to raise throughout this book involves the potential tools that a defender of a cyberattack might have at his or her disposal in order to defeat the attack in question.

One area for examination is the question of whether a suspected cyberattack comes from a human or machine opponent. In fact, this question has been at the forefront of computer science almost from its beginnings, and indeed will likely grow in its importance in the future.

Alan Turing first proposed a test, which, if it could be answered, could distinguish a third party in a dialogue as being human or a computer. This was not in the realm of a cyberattack but a random dialogue between human and machine. Turing proposed an experiment with two participants in separate rooms so they had no visual contact. In one room was a questioner who could ask any question at all of the entity in the other room, who could be a human or computer. If, after the questioner was allowed as many questions as possible of the other entity, the questioner could not determine with absolute certainty if the entity in the other room was a human or a machine, and indeed it was a machine, it could be said that the machine possessed human intelligence. Turing published this article in the journal *Mind* in 1950 (Figure 16.1).

One of the earliest initiatives to test the Turing challenge was a program called Eliza developed by Joseph Weizenbaum at MIT in 1966 (Weizenbaum, 1966). With the Eliza program, a questioner could make as many statements or questions as desired of Eliza, who was allegedly a psychotherapist. Here is a short dialogue from the original Eliza program:

```
ELIZA: Is something troubling you ?
YOU:   I'm not feeling well today.
ELIZA: Do you know anyone else who is not feeling well today ?
YOU:   No, just me.
ELIZA: Why not ?
YOU:   I have a headache.
ELIZA: You say you have a headache ?
YOU:   Yes.
ELIZA: I understand.
YOU:   What can I do to get rid of my headache.
ELIZA: Is it important to you that you have your headache ?
YOU:   Yes, I definitely need to end my headache.
ELIZA: You seem to be quite positive.
YOU:   Absolutely.
ELIZA: Does that have anything to do with the fact that you
       have your headache ?
```

The usual result of Eliza tests was for the human user to quit in frustration at the repetitiveness of Eliza's answers. No person would reach the conclusion that Eliza was actually a human.

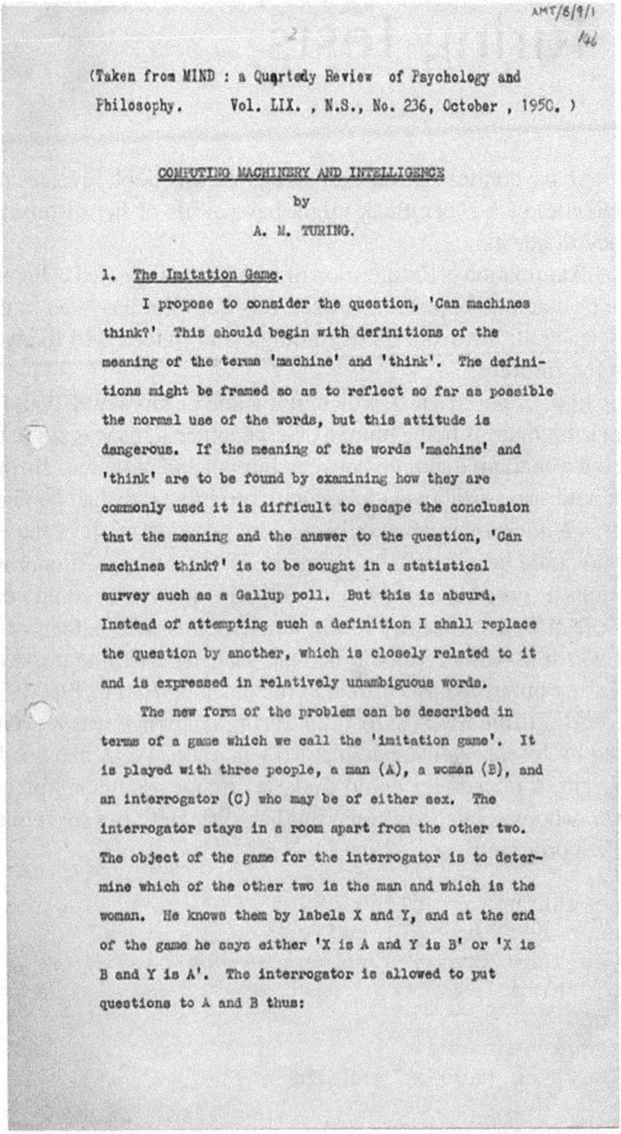

AMT/B/9/1

2

(Taken from MIND : a Quartedy Review of Psychology and Philosophy. Vol. LIX. , N.S., No. 236, October , 1950.)

COMPUTING MACHINERY AND INTELLIGENCE

by

A. M. TURING.

1. The Imitation Game.

I propose to consider the question, 'Can machines think?' This should begin with definitions of the meaning of the terms 'machine' and 'think'. The definitions might be framed so as to reflect so far as possible the normal use of the words, but this attitude is dangerous. If the meaning of the words 'machine' and 'think' are to be found by examining how they are commonly used it is difficult to escape the conclusion that the meaning and the answer to the question, 'Can machines think?' is to be sought in a statistical survey such as a Gallup poll. But this is absurd. Instead of attempting such a definition I shall replace the question by another, which is closely related to it and is expressed in relatively unambiguous words.

The new form of the problem can be described in terms of a game which we call the 'imitation game'. It is played with three people, a man (A), a woman (B), and an interrogator (C) who may be of either sex. The interrogator stays in a room apart from the other two. The object of the game for the interrogator is to determine which of the other two is the man and which is the woman. He knows them by labels X and Y, and at the end of the game he says either 'X is A and Y is B' or 'X is B and Y is A'. The interrogator is allowed to put questions to A and B thus:

FIGURE 16.1 Excerpt from Turing draft for *Mind*. (© Reproduced from *Mind* vol. LIX, Oct. 1950 with the permission of Oxford University Press. Turing Digital Archive.)

It is interesting that in the original formulation of the Turing test, Turing described the process with the questioner as indicated above, but the respondent not visible in a nearby room would be either a male or a female. Undoubtedly, Turing described the challenge in 1950 realizing that it would merely confuse his readers to imagine a computer (or an electronic machine) in another room, since at that time there were only a handful of computers in the world and very few readers would have understood the context and importance of the problem.

16.1 INTRODUCTION

In developing this field of behavioral cybersecurity, we have researched the possibility of detecting the gender of a writer, such as a hacker in a computing environment.

It has been noted in the current research that posing the Turing test challenge in the context of gender determination is in fact the manner by which Turing himself chose to explain the concept of his test to an audience that might have been challenged by the idea that machines could conduct a dialogue with human interrogatories.

16.2 THE ROLE OF THE TURING TEST IN BEHAVIORAL CYBERSECURITY

Given the confluence of external events, the power of the Internet, increasing geopolitical fears of "cyberterrorism" dating from 9/11, a greater understanding of security needs and industry, and economic projections of the enormous employment needs in cybersecurity, many universities have developed more substantial curricula in this area, and the United States National Security Agency has created a process for determining Centers of Excellence in this field (NCAEIAE, 2018).

At the 1980 summer meeting of the American Mathematics Society in Ann Arbor, Michigan, a featured speaker was the distinguished mathematician the late Peter J. Hilton (Pedersen, 2011). Dr. Hilton was known widely for his research in algebraic topology, but on that occasion he spoke publicly for the first time about his work in cryptanalysis during World War II at Hut 8 in Bletchley Park, the home of the now-famous efforts to break German encryption methods such as the Enigma.

The first author was present at that session and has often cited Professor Hilton's influence in sparking his interest in what we now call cybersecurity. Hilton at the time revealed many of the techniques used at Bletchley Park in breaking the Enigma code. However, one that was most revealing was the discovery by the British team that, contrary to the protocol, German cipher operators would send the same message twice, something akin to, "How's the weather today?" at the opening of an encryption session. (This discovery was represented in the recent Academy-Award nominated film *The Imitation Game* [Sony, 2014].) Peter Hilton was portrayed in *The Imitation Game*, only called "Peter" in the dialogue, but listed in the credits for the actor Matthew Beard in the role of "Peter Hilton." Of course, it is well known in cryptanalysis that having two different encryptions of the same message with different keys is an enormous clue in breaking a code. Thus, it is not an exaggeration to conclude that a behavioral weakness had enormous practical consequences, as the Bletchley Park teams have been credited with saving millions of lives and helping end the war.

16.3 A FINAL EXAM QUESTION

In an offering of the undergraduate course Behavioral Cybersecurity at Howard University in the spring semester 2016, there was considerable discussion about the identification of the gender of a potential hacker or other computer user. This led to a question on the final examination that asked students the following:

> We know, in general in the US as well as at Howard [at the time, the first author's university], that only about 20% of Computer Science majors are female. Furthermore, of those CS students choosing to concentrate in Cybersecurity, fewer than 10% are female. Can you suggest any reason or reasons that so many fewer female computer scientists choose Cybersecurity?

16.4 WHILE GRADING

In the course of grading this examination, the first author, as he read each answer, questioned himself as to whether he could determine if the author of the answer was one of his male or female students, based on the tone, language choice, use of certain keywords, and expected perception of the point of view of the author. Thus, as he found that this was very often difficult to determine, it seemed that it might be interesting for other persons of widely varied backgrounds—for example: gender, age, profession, geographic location, first language—to be posed the same questions.

Consequently, a test was constructed from the student responses. Three variations were added: the first author himself wrote one of the responses in an attempt to deceive readers into thinking the writer was female, and two responses were repeated in order to validate whether the responders could detect the repetition, thus showing that they were concentrating on the questions themselves.

The specific questions are listed at the end of the chapter.

16.5 TURING'S PAPER IN *MIND*

It was noted that there was a certain similarity between the administration of this test and the classic Turing test originally posed by Turing to respond to the proposition that machines (computers) could possess intelligence.

The Turing test supposes that a questioner is linked electronically to an entity in some other location, and the only link is electronic. In other words, the questioner does not know whether he or she is corresponding with a human being or a computer. The test is conducted in the following way: the questioner may ask as many questions as he or she desires, and if, at the end of the session, the questioner can absolutely determine that the invisible entity is human, and in fact it is a computer, then the other entity can reasonably be said to possess intelligence.

There has been a large body of research on this topic since Turing first posed the question around 1950, beginning with the development of the artificial "psychologist" named Eliza, originally developed by Weizenbaum (1966), and leading all the way to IBM supercomputer Watson, which was able to beat two human experts on the game show Jeopardy! in 2010 (Baker, 2012). Regarding the Jeopardy experiment, Watson was the overall winner, but on some questions where there may have been semantic interpretations, Watson simply failed, while on many others, the speed of Watson enabled it to beat the human contestants. On a more general level, it is generally accepted, however, that no computer, however powerful, has been able to pass the Turing test.

It has been discussed throughout the history of computer science whether this test has been satisfied or indeed if it could ever be satisfied.

16.6 *THE IMITATION GAME*

It seems very interesting, in the context of the gender test as described in our course, that in many ways it draws historically from Turing's thinking.

Many readers may note that the recent film, as indicated above, addressing both Turing's life and his efforts in breaking the Enigma Code in the Second World War was called *The Imitation Game.* Turing published an extremely important article in the May 1950 issue of *Mind* entitled "Computing Machinery and Intelligence." More to the point, he called the section of this paper in which he first introduced his Turing test "The Imitation Game," which evolved into the title of the biographical film.

It is significant, in our view, that in order to explain to a 1950s audience how to establish whether an entity possessed intelligence that to describe the test in terms of a human and machine would be incomprehensible to most of his audience, since in 1950 there were only a handful of computers in existence.

Consequently, in introducing the nature of his test, he described it as a way of determining gender, as follows:

> I propose to consider the question, 'Can machines think?' This should begin with definitions of the meaning of the terms 'machine' and 'think'. The definitions might be framed so as to reflect so far as possible the normal use of the words, but this attitude is dangerous. If the meaning of the words 'machine' and 'think' are to be found by examining how they are commonly used it is difficult to escape the conclusion that the meaning and the answer to the question, 'Can machines think?' is to be sought in a statistical survey such as a Gallup poll. But this is absurd. Instead of attempting such a definition I shall replace the question by another, which is closely related to it and is expressed in relatively unambiguous words.
>
> The new form of the problem can be described in terms of a game which we call the 'imitation game'. It is played with three people, a man (A), a woman (B), and an interrogator (C) who may be of either sex. The interrogator stays in a room apart from the other two. The object of the game for the interrogator is to determine which of the other two the man is and which is the woman. He knows them by labels X and Y, and at the end of the game he says either 'X is A and Y is B' or 'X is B and Y is A'. The interrogator is allowed to put questions to A and B thus:
>
> C: Will X please tell me the length of his or her hair?

We thus view the initiative that we have developed from the Behavioral Cybersecurity course as a descendant in some small way of Turing's proposition.

16.7 RESPONDENTS

In order to understand the ways in which persons interpret written text and try to assign gender to the author—in effect a version of the gender Turing test (henceforth, GTT) described by Turing in the paper cited above—a number of individuals from

TABLE 16.1
Demographics of Participants in the Gender Turing Test

Gender	Female	Male	Total
	24	31	55
Age	Older	Younger	Total
	14	41	55

Nationality	Cameroon	Caribbean	Canada	Mexico	
	10	3	1	2	
	Puerto Rico	Eastern Europe	Saudi Arabia	USA	Total
	1	2	2	34	55

Profession	Anthropology	Computer Sci	Engineering	
	1	32	4	
	Linguistics	Psychology	Student	Total
	1	5	12	55

varying backgrounds, genders, ages, first languages, countries, and professions were given the test in question.

There were 55 subjects completing this test, and a description of their demographics shown in Table 16.1.

The participants were selected as volunteers, primarily at occasions where the first author was giving a presentation. No restrictions were placed on the selection of volunteer respondents, nor was there any effort taken to balance the participation according to any demographic objective.

The voluntary subjects were (except on one occasion) given no information about the purpose of the test and were also guaranteed anonymity in the processing of the test results. There was no limit on the time to take the test, but most observed respondents seemed to complete the test in about 15 minutes.

16.8 SUMMARY OF RESULTS

The responses were scored in two ways. First, the number of correct answers identifying the student author was divided by the total number of questions (24) in the complete test. Alternatively, the score was determined by the number of attempts. Since in only two of 19 instances the difference between the two exceeded 2%, it was decided to use the second set of response scores, which are presented in Table 16.2.

Observations of the results of these responses from the 55 participants in this study and the very diverse experiences that they brought to the response to this test yield some very interesting questions to ponder.

First, female respondents were more accurate in the identification of the gender of the students by a margin of 56.89% to 51.02%.

TABLE 16.2
Responses to the Gender Turing Test Questions

Respondents by Gender	Female	Male
Correct percentage of responses	56.89%	51.02%

Respondents by Age	Older	Younger
Correct percentage of responses	57.60%	51.77%

Respondents by Nation	Cameroon	Caribbean	Canada	Mexico
Correct %	49.17%	56.67%	45.83%	54.35%
	Puerto Rico	Eastern Europe	Saudi Arabia	United States
Correct %	50.00%	66.67%	33.33%	54.79%

Respondents by Profession	Anthropology	Computer Sci	Engineer
Correct %	64.71%	52.91%	58.33%
	Linguist	Psychologist	Student
Correct %	66.67%	54.17%	50.03%

Next, older respondents were more accurate in their identification than younger responses by a similar margin of 57.6% to 51.77%. This might be a more surprising result since for the most part the older respondents were not as technically experienced in computer science or cybersecurity matters than the younger responders, who for the most part were students themselves.

One very clear difference is that Eastern European respondents scored far higher in their correct identification of the students' gender, averaging 66.67%, with the nearest other regional responses being fully 10% less. The number of respondents from Eastern Europe was very small, so generalizations might be risky in this regard. However, the Eastern Europeans (from Romania and Russia) were not first-language speakers of English, although they were also quite fluent in the English language. Each of them also tied for the highest percentage of correct answers of anyone among all 55 respondents.

There were a fairly large number of Spanish-speaking respondents, and a number of them were not very fluent in English. Nevertheless, the overall score of the Spanish-speaking respondents was above the average for all respondents from English-speaking countries—including Cameroon, the Caribbean, Canada, and the United States. Both Cameroon and Canada are bilingual French- and English-speaking countries, but all of the respondents in this case were from the English-speaking parts of these two countries. In addition, the Caribbean respondents were also from the English-speaking Caribbean.

The Saudi Arabian respondents, for whom Arabic was their first language, had greater difficulty in identifying the correct gender. It is possible that these differences could have arisen from the lack of fluency of these respondents in English.

Of the respondents from the various disciplines, the linguist, anthropologist, engineers, and psychologists all fared better than the computer scientists—and lowest of all were the students who took the test (as opposed to the students who wrote the original answers).

It is possible, of course, to view the entire data set of responses to this test as a matrix of dimensions 24 × 55, wherein the students who wrote the original exam—and thus in effect, created the GTT—represent the rows of the matrix, and the gender classifications by the 55 responders are the columns. If we instead examine the matrix in a row-wise fashion, we learn of the writing styles of the original test takers and their ability (although inadvertent, because no one, other than the first author, planned that the writings would be used to identify the gender of the writer).

Thus, it is perhaps more informative than the assessment of the ability of the respondent to determine the gender of the test takers to note that several of the original test takers were able, unconsciously, to deceive over two-thirds of the respondents. Fully one-quarter (6 of 24) of the students reached the level of greater than two-thirds deception. Of these six "high deceivers," three were female and three were male students.

At the other end of the spectrum, one-third of the students were not very capable of deception—fooling less than one-third of the respondents. Of these eight students, six were male and only two were female. On the whole, averaging the level of deception by the male and female students, on average the female students were able to deceive 52.5% of the respondents, while the male students were only able to accomplish this with 42.2% of the respondents. The following chart shows a scatter plot of the student takers' ability to fool the respondents (Figures 16.2 through 16.4).

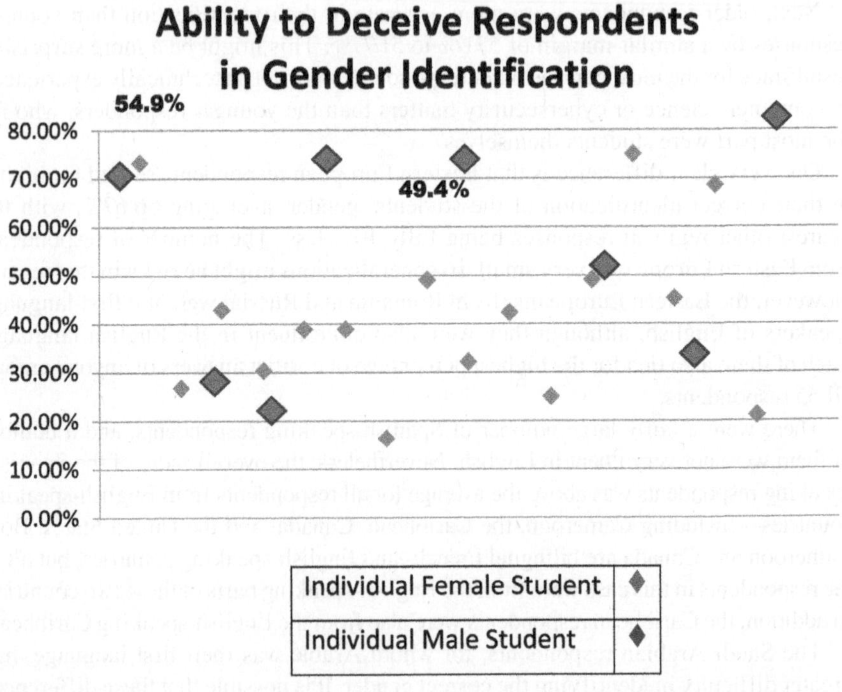

FIGURE 16.2 Scatterplot of female and male students' success in deceiving respondents.

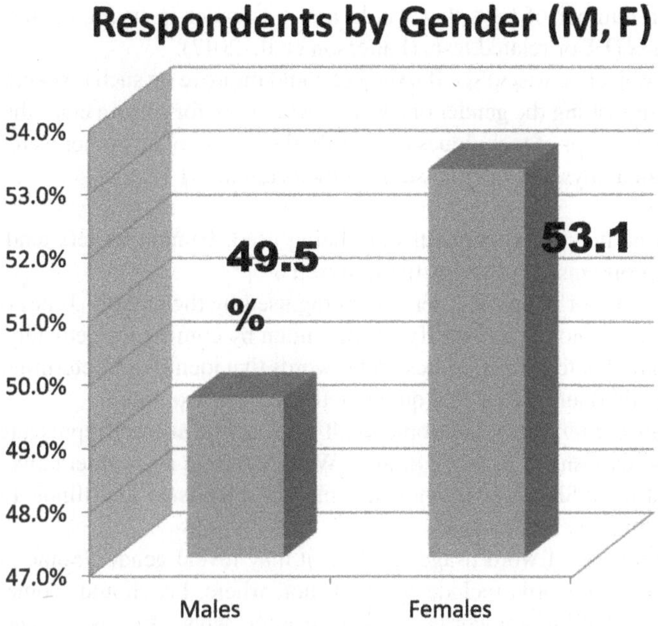

FIGURE 16.3 Respondents by gender.

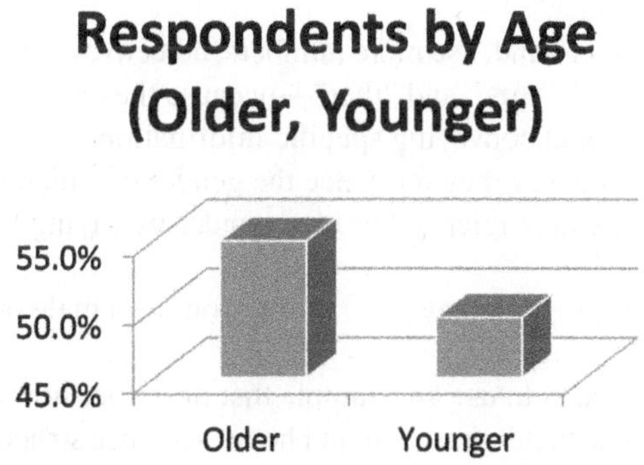

FIGURE 16.4 Respondents by age.

16.9 "COACHING" RESPONDENTS

All of the respondents described above had simply been given a test with only the simple instruction described in the attachment, without any prior preparation or understanding on the part of the respondent as to possible techniques for identifying the gender of a writer or author.

Consequently, we determined that it would be useful to see if persons could be given some training in order to try to improve their results on the GTT. We attempted to identify a number of keys that would assist a reader in trying to improve their scores on the GTT or related tests (Patterson et al., 2017).

Our next objective was to see if a subject could improve on such text analysis in the case of distinguishing the gender of a writer by looking for certain clues that could be described. A number of techniques to identify the gender of an author were described to perform an analysis of the questions in the original GTT:

1. Examine how many pronouns are being used. Female writers tend to use more pronouns (I, you, she, their, myself).
2. What types of noun modifiers are being used by the author? Types of noun modifiers: a noun can modify another noun by coming immediately before the noun that follows it. Males prefer words that identify or determine nouns (a, the, that) and words that quantify them (one, two, more).
3. *Subject matter/style*: The topic dealt with or the subject represented in a debate, exposition, or work of art. "Women have a more interactive style," according to Shlomo Argamon, a computer scientist at the Illinois Institute of Technology in Chicago.
4. Be cognizant of word usage and how it may reveal gender. Some possible feminine keywords include: with, if, not, where, be, should. Some of the other masculine keywords include: around, what, are, as, it, said. This suggests that language tends to encode gender in very subtle ways.
5. "Women tend to have a more interactive style," said Shlomo Argamon, a computer scientist at the Illinois Institute of Technology in Chicago (Argamon et al., 2003). "They want to create a relationship between the writer and the reader."

   Men, on the other hand, use more numbers, adjectives, and determiners—words such as "the," "this," and "that"—because they apparently care more than women do about conveying specific information.
6. Pay attention to the way they reference the gender of which they speak. For example, a female may refer to her own gender by saying "woman" rather than "girl."
7. Look at the examples that they give. Would you see a male or female saying this phrase?
8. A male is more likely to use an example that describes how a male feels.
9. Women tend to use better grammar and better sentence structure than males.
10. When a person of one gender is describing the feelings/thoughts of the opposite gender, they tend to draw conclusions that make sense to them, but will not provide actual data.

It should be noted that some prior work includes the development of an application available on the Internet (Gender Guesser), developed by Neil Krawetz based on Krawetz (2018) and described at the location http://hackerfactor.com/GenderGuesser.php.

This application seems to depend on the length of the text being analyzed, and in comparison to the responses of our human responders, does not perform as well, as normally the application indicates the text is too short to give a successful determination of gender.

However, because the overall objective of this research is to determine if a GTT can be used in a cybersecurity context, it is likely that an attacker or hacker might only be providing very short messages—for example, a troll on the Internet trying to mask his or her identity in order to build a relationship, say, with an underage potential victim.

16.10 FUTURE RESEARCH

The questions that have been raised by this research have also opened the potential of devising other such tests to determine other characteristics of an author, such as age, profession, geographic origin, or first language. In addition, given that the initial respondents to the test as described above are themselves from a wide variety of areas of expertise, nationality, and first language, a number of the prior participants have indicated interest in participating in future research in any of these aforementioned areas.

REFERENCES

Argamon, S., Koppel, M., Fine, J., and Shimoni, A. R. 2003. Gender, genre, and writing style in formal written texts. *Text*, 23(3), 321–346.

Baker, S. 2012. *Final Jeopardy: The Story of Watson, the Computer That Will Transform Our World*. Boston: Houghton Mifflin Harcourt.

Krawetz, N. 2018. Gender Guesser, http://hackerfactor.com/GenderGuesser.php

National Centers of Academic Excellence in Information Assurance Education (NCAEIAE). 2018. National Security Agency. https://www.nsa.gov/ia/academic_outreach/nat_cae/index.shtml

Patterson, W., Boboye, J., Hall, S., and Hornbuckle, M. 2017. The Gender Turing Test, *Proceedings of the AHFE 2017 International Conference on Human Factors in Cybersecurity*, (593), 281–289.

Pedersen, J. (ed.). 2011. Peter Hilton: Codebreaker and mathematician (1923–2010). *Notices of the American Mathematics Society*, 58(11), 1538–1551.

Sony Pictures Releasing. 2014. *The Imitation Game* (film).

Turing, A. M. 1950. Computing machinery and intelligence. *Mind: A Quarterly Review of Psychology and Philosophy*, LIX(236), 433–460.

Weizenbaum, J. 1966. ELIZA—A computer program for the study of natural language communication between man and machine. *Communications of the Association for Computing Machinery*, 9, 36–45.

PROBLEMS

16.1 Read Turing's article in *Mind*. How would a reader respond (a) in 1950; (b) today?

16.2 Try Eliza. How many questions did you ask before Eliza repeated?

16.3 Construct three questions for Jeopardy! that the human contestants would probably answer faster or more correctly that Watson.

16.4 Give examples of two encryptions of the same message with different keys. Use the encryption method of your choice.

16.5 Consider developing your own gender Turing test. List five sample questions that might differentiate between a female or male respondent.

16.6 Consider developing an age Turing test. List five sample questions that might differentiate between a younger or older respondent.

16.7 Take the "Who Answered These Questions" gender Turing test. To find out your score, email your responses to waynep97@gmail.com. Send a two-line email. The first line will have "F: x1, 2, x3, …" and the second line "M: y1, y, y3, …." You will receive an email response.

16.8 Comment in the context of 2019, from Turing's article in *Mind*, on interrogator C's question in order to determine the gender of the hidden subject.

16.9 Run the Gender Guesser on (A) your own writing and (B) the "Test Bed for the Questionnaire" below.

Test Bed for the Questionnaire: "Who Answered This Question?"

A group of cybersecurity students were asked, "We know, in general in the US as well as at Howard, that only about 20% of Computer Science majors are female. Furthermore, of those CS students choosing to concentrate in Cybersecurity, fewer than 10% are female. Can you suggest any reason or reasons that so many fewer female computer scientists choose Cybersecurity?"

Do you think the respondents in lines 1–24 are female (enter F) or male (enter M)?

No.	Response
1	A few reasons so many fewer female computer scientists choose Cybersecurity. There is a lack of hiring opportunities for women in cybersecurity. The levels of pay for both men and women doing the same job are uneven. Women are not as interested in hacking which studies show is "more appealing to men"-The increasingly dominant way in which competitive hackathons in computer science education, training and recruiting are used which deter a lot of women. Many women are afraid to step out of their comfort zones. Women might not see other women in the cybersecurity field succeeding as they planned which might deter them. The number of women that are taking up technology subjects, such as computer science is at an all time low.
2	Sexism still exists in our society and girls find facing this at a young age. Think about the toy aisle with its distinct pink and blue color coding. The message that the toy aisle often sends is that girls are meant to be homemakers, caretakers and nurturing while boys are supposed to go places, design things and build stuff. Also I believe the stigma of women not being as versed in mathematic must cease if we would like to lessen the divide in gender gaps. We have to combat the cultural belief that some people are simply born with math, science or computer talent and others are simply "not good at it." There's lots of research that shows that persistence and hard work play a much larger role in success in any area of science and engineering than "native ability." All these key factors compound into giving reason why Cybersecurity features so little women.
3	Well I think it is because most people think that cyber security is all about hacking and stopping hacking, which is somewhat true but they also thin that hackers are people are always behind their computers and coding in 0 s and 1 s like machine language and so to stop them would mean that a cyber security person would then have to always up on their computers and know machine language like the back of their hands and most people don't think that to be fun. I also think it could be because of the fact that we are always told that cyber security is a lot of math and logics like that and generally not a lot of females enjoy to math and logic stuff so again they won't find it interesting. So I think it is really down to people not really knowing what cyber security covers and that there are many aspects of it that people have no clue about and the stereotype that is known is not something that generally interests a lot of females.

(*Continued*)

No.	Response
4	One reason that comes to my mind is possible gender discrimination. Similar to the Navy recently allowing women to work on submarines, this particular sector of work is mostly dominated by men and has been for many years. I think that women are skeptical to major in this field because of the fear of being competitive with men as well as possibly being looked at strangely by men who may wonder why a female is interested in this subject. Additionally, since there are so few females in this particular field some may be afraid because there are not many women to talk about Cybersecurity, or to follow/compare in terms of success.
5	I believe that both males and females are given equal incentives to concentrate in Cybersecurity. Cybersecurity is a young and burgeoning industry which in the last 20 years with the advancement of technology is spawning a number of secure career opportunities. I think that successful female Cybersecurity experts, like ambassadors, can do a lot to promote and encourage females in Cybersecurity.
6	I believe men are by nature more aggressive and therefore more easily attracted to cybersecurity because of the terminology widely used in introducing the field, like cyber warfare, hacking, malicious behavior, etc. Women who persist and get beyond these initial deterrents learned that the actual issues in cyber security are more related to the careful examination of facts and events that allow for success in defending our electronic environments.
7	In my opinion, there are a few reasons. I think 1 of the main reasons is the same general reason most women aren't interested in Computer Science: it's a male-dominated area which often means the workplace is cultivated by practices/behavior/feelings that are exclusive, discriminatory, and often demeaning towards women. These practices, behaviors, and feelings are definitely not restricted to Computer Science and in fact are present in workplaces across the world, however in areas where the ratio of women to men is so imbalanced as it is in Computer Science and Cybersecurity, it tends to run rampant. Women often have to go above and beyond to prove their competence, capabilities, and that they belong. Still common in the general workplace, they also continue to fight for equal pay, even when just as qualified, if not more qualified, than their male counterparts. Aside from dealings in the workplace, men like to hack/break into things and make video games, whereas women tend to be more interested in creating, designing, and building things (other than video games).
8	I personally believe that women are not in the cyber security field for three separate reasons. Number one, security fields of any sort are usually male dominated and don't have a large amount of women involvement. From security guards, to policeman, and even in the military males dominate the industries. With that being said it's not necessarily a natural inclination for women to be attracted to security industries nonetheless cyber security. Number two; computer science in general is a very secluded field whereas many women want to feel included and more "social" while at work. The nature of cyber security comes off as a heavily coding field with tons of hackathons, and sleepless nights with smelly men in hot computer labs. Even though this isn't the case it's not the most attractive field for women to want to step into. Lastly, number three; everyone wants to work with people who are like them. Men don't make it easy for women to assimilate into the field. This can discourage some women from the outside looking in, who don't have anyone to relate to.
9	Cybersecurity might be unappealing/or less appealing to ladies because it seems to be a cutthroat, slugitout, good guy vs bad guy field with constant mental and technical battles over the course of days or weeks. This is not to say that ladies are averse towards battle, (far from it), it just seems that tradition and culture tend to encourage ladies to explore more flamboyant and outwardly creative fields as opposed to technically creative fields.

(*Continued*)

No.	Response
10	I feel as if cybersecurity is a field that is not introduced to many incoming computer science students in general. If there was more exposure to cybersecurity in intro courses there would be an increase of students, including females, choosing to concentrate in cybersecurity.
11	I think the reason why so few female computer scientists choose cybersecurity is because the word cybersecurity has a connotation of hacking, and hacking is something that is very risky. Females are more risk-averse than men.
12	I believe that most females would rather create software and pursue careers as that would allow this rather than focus on cyber security.
13	I think the ultimately the computer science field is very competitive and it doesn't really get advertised a lot as a field for women. When people hear engineering, not many people think of women as engineers so it becomes a little discouraging for a woman to want to go into this career field. Also, when it comes to resume building, a lot of people look for things like hackathons for experience. In my personal opinion, hackathons are fun, but not something that I would want to take part in repeatedly. Having to stay up for over 24 hours, sitting at tables in front of my computer, and not bathing; that just isn't an environment that appeals to women.
14	I do believe that most don't know much about Cybersecurity. Even those in computer science, the knowledge is limited to those who venture into that facet of computer science. So, there is limited number of females who know much about CS, there is even less who know much about Cybersecurity. What they don't know they won't even have a chance to be interested which would explain the low percentage of females interested in Cybersecurity. Generally, I believe the percentage of females that do know a little bit about Cybersecurity may find that little bit boring.
15	Personally, I do not think the cyber security field is that appealing to the females in the computer science field. The stereotype that often comes with cybersecurity is the "hacker." The same hacker who stays up for nights, living in a basement, writing code, often trying to do something illegal (at least through the stereotype). I could see how this stereotype alone could repel women from the field. Additionally, cybersecurity is not the "sexiest" form of computer science. There's not a lot of UI design or code that involves in making something visually appealing, such as a website or an app.
16	There is a stigma surrounding women in technology & engineering disciplines and it can be spread directly and indirectly. I've personally spoken to some women that majored in Computer Science and they shared with me their experiences of sexism. They felt encouraged to enroll in the "non-coding" by faculty members, as if they didn't have the competency to take coding intensive courses. In this situation, the person was unaware of their bias towards women in technology. I think the leadership of departments that have underrepresentation of women should enroll the faculty in unconscious bias training, because they could be inadvertently discouraging some of the brightest minds away from the field.
17	I firmly believe that the low percentage of females in cybersecurity is due to the oppression of women in society. It is a known fact that women were not even allowed to vote just over a hundred years ago. Women have had to make many strides, especially in the work field and unfortunately engineering, specifically computer science is a field many women are discouraged to engage. Women are more so seen in positions of secretaries, nurses, and other fields that are seen as less important because society believes men are more applicable for the positions. Women on average even make less money than men and it is even a greater difference for African American and Latino Women.

(*Continued*)

No.	Response
18	I think it's related to human mindset we have towards women being the lesser gender and computer science and cybersecurity being a "field for men." With the women going into other "male dominated" fields like construction and finance increasing, I think it will also improve with cybersecurity and computer science.
19	(1) Media representation of cybersecurity professionals are not female friendly and hence does not initiate interest in female in cybersecurity. (2) There exists a level of difference in the pay rate of a male Cybersecurity professional and female Cybersecurity professional, hence discouraging females to join the industry. (3) A study showed that the basic instinct of females have an inclination towards creativity and designing. Hence, most women computer scientists choose UX design to any other field inside computer science.
20	As far as STEM in general, I think women are not represented as well as men because of the sexism in the field. I think female computer scientists choose other concentrations over cyber security because it is an emerging field. So men have been working in computer science longer and are more likely to have been exposed to cyber security. I do not think that is the case for women, who for the most part, are starting to be exposed to STEM majors and computer science.
21	Historically, women were not encouraged to pursue STEM related fields such as engineering and computer science. Often times, they are prompted to go into education or careers that are not as technically inclined. Additionally, there is the stereotype that all computer scientists and engineers are men. Consequently, this would cause for less women to even think about computer science or Cybersecurity. Furthermore, in regard to Cybersecurity, a majority of women are not even informed to what it actually is! Being unknowing of a topic is a good excuse to not pursue it. In the future, schools and the media should work toward increasing exposure of Cybersecurity to women rather than excluding women and teaching them that it is too difficult.
22	I think the ultimately the computer science field is very competitive and it doesn't really get advertised a lot as a field for women. When people hear engineering, not many people think of women as engineers so it becomes a little discouraging for a woman to want to go into this career field. Also, when it comes to resume building, a lot of people look for things like hackathons for experience. In my personal opinion, hackathons are fun, but not something that I would want to take part in repeatedly. Having to stay up for over 24 hours, sitting at tables in front of my computer, and not bathing; that just isn't an environment that appeals to women.
23	If you were to look at the statistics about how many computer scientist specialize in cyber security, they would also be low. Cyber is not a glamorous field. One reason why cybersecurity is not popular is because it is difficult. Another is because cyber security does not get the same attention that app development does. These reasons explain why computer science is not popular with women.
24	I do believe that most don't know much about Cybersecurity. Even those in computer science, the knowledge is limited to those who venture into that facet of computer science. So, there is limited number of females who know much about CS, there is even less who know much about Cybersecurity. What they don't know they won't even have a chance to be interested which would explain the low percentage of females interested in Cybersecurity. Generally, I believe the percentage of females that do know a little bit about Cybersecurity may find that little bit boring.

17 Personality Tests, Methods, and Assessment

The personal data of approximately 87 million Facebook users was acquired via the 270,000 Facebook users who used a Facebook app called "This Is Your Digital Life." As a result of users giving this third-party app permission to collect their data, they were also giving the app access to information on using their friends data. Underlying this test was a personality assessment of the common Big Five personality traits. This case revealed multiple aspects of interesting questions about personality tests, methods, and assessment. For example, how can an individual's personality traits be measured through a social media application in a way that yields valid and reliable personality data that can be applied to impact advertising strategies and politics? In a hypothetical world, how can the possible motivational dynamics of the researchers, businesspeople, and social media company be assessed using personality test, methods, and assessment?

Within the field of behavioral science, there are a range of research questions that researchers pursue and consider in designing research to understand human behavior. There is a research process in behavioral science that engages multiple types of considerations about the nature of what counts as knowledge (epistemology), the general approach for carrying out the research (methodology), and procedures for collecting and analyzing data (methods). The theoretical diversity in conceptualization of human personality discussed in previous chapters is matched by the methodological variation used to study personality.

The most common research designs used to explore personality are correlational, experimental, and narrative designs. Within personality psychology, there are different levels of analysis that correspond to trying to understanding how a person is like all others (i.e., human nature), like some others (i.e., individual differences), and like no others (i.e., human uniqueness) (Kluckhohn and Murray, 1953). The purpose of this chapter is to describe methods of inquiry used in personality psychology to advance the argument that personality research methods of inquiry and personality tests used to assess personality can be used to advance the science of cybersecurity by applying and generating personality research designs, questions, and methods to better understand, protect, and defend information and systems in cybersecurity situations.

17.1 RESEARCH DESIGNS USED IN PERSONALITY PSYCHOLOGY

Within behavioral science, research design is the general approach to inquiry about a phenomenon. Crotty (1998) in his book, *The Foundations of Social Research: Meaning and Perspective in the Research Process*, described a basic approach to research design in terms of four key questions that are pertinent for design: (1) What methods do we propose? (2) What methodology governs our choice and use of methods? (3) What theoretical perspective lies behind the methodology in question? (4) What epistemology informs this theoretical perspective? Crotty (1998) defined the key concepts within these questions as follows.

17.1.1 The Research Process

Methodologies (p. 7): Design strategies of inquiry (plan of action that shapes our choice and use of particular methods and links them to a desired outcome)

Methods (p. 6): Procedures and techniques to gather and analyze data (the specifics and details of carrying out our research activities/design/plan of action)

Theoretical perspectives (p. 7): Philosophical stances on the way of looking at the world and making sense of it (undergirds our choice of methodology, provides context for process and ground its logic, makes assumptions we bring to our methodology explicit: How do we know what we know?)

Epistemologies (p. 8): Theories on the nature of the "knowing" (What is the nature of knowledge?—e.g., Is it something to be discovered or constructed?)

17.1.2 Research Designs

In the following description, key distinguishing features among the most common designs, as well as special cross-cutting designs used in behavioral science to understand personality, are highlighted.

17.1.2.1 Experimental Design

- Emphasis on central tendencies (vs variation).
- Purpose of inquiry is to assess causal impact of one or more experimental manipulations on a dependent variable and draw conclusions about the causal relationships among variables.
- In essence, within experimental designs, "differences of means resulting from differences in experimental conditions are thought to reflect the direct casual effects of the independent variables on the dependent variables" (Revelle, 2007, p. 38).
- An example of a personality study using an experimental design is the Revelle et al. (1980) experimental study that explored the following research question: Is there an interactive effect of personality and time of day on caffeine?

17.1.2.2 Correlational Design

- Emphases in these designs are variability, relationship, and individual differences.
- Most common approach in personality research (Robins et al., 2007).
- Purpose of inquiry is to assess the relationships between and among two or more variables and making of predictions. However, causal relationships cannot be determined.

- An example of a personality study using correlation design is Lester et al. (2006) study that used a correlational design to explore the following research question: Is there a relationship between various dimensions of personality and use of eBay?

17.1.2.3 Narrative Design

- Emphasis on storied nature of experience (Winston, 2011; Winston-Proctor, 2018).
- Purpose of inquiry is to advance understanding of the nature of experience in terms of narrative structure, content, and performance (e.g., what individuals are trying to do with their discourse).
- Narrative approaches to personality seen in the early days of psychology, but have become increasingly popular (particularly among graduate students) in the 2000s (Singer, 2004).
- Narrative design could be considered a type of hybrid design with many variations, including psychobiography, collective comparison, case studies, longitudinal design (e.g., to study narrative identity continuity).
- An example of a personality study that uses a narrative design is McAdams et al. (2006) to answer the following research question: Do college students demonstrate continuity from freshman year to senior year in the narrative complexity and themes of agency and growth within their narrative identity?

17.1.2.4 Special Cross-Cutting Designs Used in Personality Psychology

17.1.2.4.1 Case Study Design

- Emphasis on a case or collection of cases.
- Case is a bounded system; most typically in personality psychology, the individual or the life is the bounded system.
- Purpose of inquiry ranges from theory testing to theory development and can include an idiographic or nomothetic focus. McAdams (2006) described the differences between these two approaches in the following way: personality psychologists develop and validate ways of measuring individual differences, necessitating a quantitative and focused inquiry into single dimensions of human variation within large samples of individuals—what Gordon Allport called the *nomothetic* approach to personality research. At the same time, personality psychologists aim to put the many different conceptualizations and findings about many different dimensions of human variation together into illuminating personological portraits of the individual case—what Allport called the *idiographic* approach. How to reconcile the different demands of analytic, quantitative, nomothetic studies on the one hand and synthetic, qualitative, idiographic inquiries on the other has been a central conundrum for "personality psychology since the very beginning" p. 13.
- Typically uses multiple data sources (e.g., primary documents, interviews, observations, questionnaires etc.).
- Can be a quantitative, qualitative, or mixed-methods design (e.g., McAdams and West, 1997).

17.1.2.4.2 Experience Sampling Method

- Emphasis is on measuring participants' feelings, thoughts, actions, and/or activities within context in the moment as they go about living their daily lives (Larson and Csikszentmihalyi, 1983).
- Purpose of inquiry is typically to achieve one of the following: examine individual differences in temporal and behavioral distributions, situation–behavior, contingencies, daily processes, and the structure of daily experience (Conner et al., 2009).
- Identification of patterns of behavior within a given individual, rather than strictly identifying patterns of behavior across individuals, as is typical within other standard nomothetic approaches.
- In essence, "by capturing experience, affect, and action in the moment and with repeated measures, experience sampling method [ESM] approaches allow researchers access to expand the areas and aspects of participants' experiences they can investigate and describe and to better understand how people and contexts shape these experiences" Zirkel et al. (2015).

17.2 PERSONALITY TEST, METHODS, AND ASSESSMENTS

17.2.1 Personality Trait Assessments

17.2.1.1 NEO PI-R

The NEO PI-R is a standardized questionnaire to assess the five-factor model (FFM) of personality traits (i.e., the Big Five model of personality traits) (McCrae and Costa, 1997, 1999; McCrae, 1992). The measure includes 240 items. It measures an individual's emotional, interpersonal, experiential, attitudinal, and motivational styles within the five major domains of personality traits within the Big Five model of personality traits: extroversion, neuroticism, openness, conscientiousness, and agreeableness. The individual responds to each item on a 5-point scale that ranges from 1, Strongly Disagree, to 5, Strongly Agree. It also includes assessment of the six traits (i.e., facets) of which each domain is composed. The procedure to complete the NEO-PI-R can include self-report assessment and or observer report assessment from the individual's peer, spouse, supervisor, etc. Both the self-report assessment and the observer report assessment take about 30–40 minutes to complete.

17.2.1.2 The Big Five Inventory

The Big Five Inventory (BFI) is a standardized questionnaire designed to assess the five-factor model of personality traits (i.e., the Big Five model of personality traits). Unlike the NEO-PI-R, it only includes 44 items and thus only measures each of the five major domains of personality traits rather than also including the six traits within each domain. The response scale ranges from 1, Strongly Disagree, to 5, Strongly Agree. Unlike the NEO-PI, there is a version available for public use.

17.2.1.3 Myers-Briggs Type Indicator

Though most personality psychologists do not consider the Myers-Briggs Type Indicator (MBTI) (Briggs and Myers, 1976) a personality trait assessment per se, it is often considered one by those who use it outside of academic research. The MBTI is used to create a typology of personality characteristics based on Jung's (1971) theory of personality types. The personality types that are assessed using a four-dimension score based on 95 of 166 items. The four MBTI dimensions are as follows: Extraversion–Introversion (E–I), Sensing–Intuition (S–N), Thinking–Feeling (T–F), and Judging–Perceiving (J–P).

Bess and Harvey (2002) highlighted that the MBTI is frequently used in applied organizational and assessment settings. Within these settings, it is used for hiring (i.e., employment selection and hiring), career counseling, self-development, and popular assessment reliability (e.g., Briggs and Myers, 1976; McCrae and Costa,1989; Myers and McCaulley, 1985). Even though the MBTI is frequently used in these settings, most academic psychologists regard the MBTI as a very bad measure because of its low reliability and validity (Reynierse, 2000).

17.2.2 Motivation and Personalized Goal Assessments

Within the field of psychology, there is a diversity of measures of goals and motivations developed over time (see Mayer et al., 2007, for a review of motivation measures from 1930–2005). There are several that are commonly used within personality psychology to understand a person's motivation. They can be grouped by the general approach to data gathering and analysis.

Projective Tests and Thematic Content Coding Systems: There are various projective tests that are used by psychologists to measure motivation. Among the types of motives that these tests measure are achievement, power, affiliation (social motive), and intimacy motives. Typically, these tests use thematic content coding systems to interpret these motives. Projective tests are considered by personality psychologists as indirect or implicit measures of motivation.

Thematic Apperception Test (Atkinson, 1958; McAdams, 1984; Morgan and Murray, 1935; Murray, 1938; Winter, 1973; Woike et al., 1999). This test has prompts to which the individual responds to tell a story about a picture. Then the researcher interprets these stories, most often using a coding system that has been developed by psychologists as valid and reliable across studies. Examples of these coding systems are published in Smith (1992) as follows: the intimacy motivation scoring system (McAdams, 1980, 1992), the affiliation motive scoring manual, the affiliative trust-mistrust scoring system (McKay, 1992), the achievement motive scoring manual (McClelland., 1958), the scoring manual for the motive to avoid success (Horner and Fleming, 1977), and the scoring system for the power motive (Winter, 1992).

Achievement motivation questionnaires and scales are used to measure an individual's motivation. With the exception of the personality research form (Jackson et al., 1996), almost all of these are designed to measure the achievement motivation of students who are at various levels of their development (i.e., children, adolescents, and young adults).

17.2.2.1 Personality Research Form

This test measures achievement motivation along with 21 other types of motives.

As Mayer et al. (2007) indicated, the Personality Research Form (Jackson et al., 1996) is the second most common test of general motivation, with the Thematic Apperception Test (TAT) being the most common.

Academic Motivation Scale (AMS) (Vallerand et al., 1992)

This scale assesses an individual's achievement motivation locus along with other subscales that measure intrinsic and extrinsic motivation.

Motivated Strategies for Learning Questionnaire (MSLQ) (Pintrich et al., 1993)

This questionnaire includes 15 scales of strategies for college students measuring cognitive and motivational components of learning and resource management.

Personalized Goals

Measures of personalized goals are considered by personality psychologists as direct or explicit measures of goals.

17.2.2.2 Personal Project Assessment

Personal project analysis (Little, 1983) assesses the kinds of activities and concerns that people have over the course of their lives. Individuals can have a number of personal projects at any given time that they think about, plan for, carry out, and sometimes (though not always) complete. Using this assessment, the individual is first asked to do the following: write down as many personal projects and activities as he/she can in which he/she is currently engaged in or considering. This is followed by the individual being asked to select 10 projects that are important to how he/she feels and then asked to rate each on the following dimensions: importance, difficulty, visibility, control, responsibilities, time adequacy, outcome likelihood of success, self-identity, others' view of importance, value congruency, progress, challenge, absorption, support, competence, autonomy, stage, and feelings. With these components in combination, psychologists use various matrices for interpretation of the individuals' patterns of personal goals.

17.2.2.3 Personal Strivings Assessment

The Personal Strivings Assessment (Emmons, 1986) contains 15 items beginning with "I typically try to…" After the individual finishes writing down his/her strivings, the researcher provides him/her with the strivings assessment matrix. On this matrix, the individual rates each of the 15 strivings on the following dimensions: the degree of commitment/investment in the goal (commitment/intensity), the degree to which the goal is perceived as stressful/challenging (ease/effort), and the anticipated outcome/reward (desirability/reward) (Emmons, 1989). Another way to interpret the personal goals from the list of 15 strivings the individual generates is to analyze their motivational themes. Emmons (1989), developed a coding manual for identifying 12 oerarching motivational themes across the person's set of strivings: avoidance goals, achievement, affiliation, intimacy, power, personal growth and health, self-presentation, independence, self-defeating, emotionality, generativity, and spirituality.

17.2.3 Narrative Personality Assessments

17.2.3.1 Psychobiography

Psychobiographical approaches to assessing human personality focus on a single individual, though uses and applications of psychobiography vary. For example, Elms (2007) identified the following uses of psychobiography within personality psychology: understanding of unique personality processes, clinical diagnosis, predicting political candidates' performance in office, sources of hypothesis and theories, validation or invalidation of hypotheses and theories. Though the focus of psychobiography historically and more recently has been on individuals (e.g., Elms and Heller, 2005; McAdams and West, 1997; Runyan, 2005; Schultz, 1999; Stewart et al., 1988), there are some examples of psychobiography that focus on multiple individuals in the form of comparative case studies. Examples of psychobiographies are Alexander's (1990) comparison of Freud and Jung, Elms's (1986) comparison of two national leaders and four foreign policy advisors, Atwood and Stolorow's (1993) examination of four personality theorists, and Nasby and Read (1997) case study of the personality of Dodge Morgan.

The life story interview (McAdams, 1985) is a semistructured oral interview instrument that captures the story of a person's life. With this personality instrument, narrative identity constructs are assessed through interview probes that give the respondent the opportunity to describe key scenes, characters, and plots in the stories of their lives. More specifically, McAdams (2007) describes the life story interview as a 2-hour procedure where an individual provides a narrative account of his or her life—past, present, and imagined future—by responding to a series of open-ended questions. The procedure begins by asking the respondent to divide his or her life into chapters and provide a brief plot outline of each. Next, the interview asks for detailed accounts of eight key scenes in the story, including a high point, a low point, and turning point scene. The interview protocol also includes imagined future chapters and the basic values and beliefs on which the story's plot is developed.

17.2.3.2 The Guided Autobiography Instrument

The guided autobiography (McAdams, 1997) is a written structured instrument that is organized around the idea of critical life events or episodes. McAdams (1997) described an episode as a specific happening that occurs in a particular time and place that stands out for some reason. The procedure for this instrument centers on asking the participant to write down a description of at least a paragraph or two for each of the following critical life events or episodes: peak (high point), nadir (low point), turning point, continuity, childhood, adolescent, morality, goal.

For each episode, the procedure asks the individual to think about the event carefully and include all of the following in their written description of the event:

1. When did the event occur? (How old were you?)
2. What exactly happened in the event?
3. Who was involved in the event?
4. What were you thinking, feeling, and wanting in the event?
5. Why do you think that this is an important event in your life story? What does this event say about who you are, who you were, who you might be, and how you have developed over time?

17.2.3.3 The Self-Defining Memory Task

The self-defining memory task (Singer and Blagov, 2004) is an instrument that captures self-defining memories. A self-defining memory is a personal memory that includes very specific attributes. The self-defining memory task instrument asks the respondent to recall a personal memory that includes the following attributes: (1) is at least 1 year old; (2) is a memory from their life that they remember very clearly and that still feels important to them even as they think about it; (3) is a memory about an important enduring theme, issue, or conflict from their life; (4) is a memory that helps explain who they are as an individual and might be the memory they would tell someone else if they wanted that person to understand them in a profound way; (5) is a memory linked to other similar memories that share the same theme or concern; (6) may be a memory that is positive or negative, or both, in how it makes them feel. The only important aspect is that it leads to strong feelings; (7) is a memory that they have thought about many times. It should be familiar to them like a picture you have studied or a song (happy or sad) you have learned by heart.

The procedure begins by asking the respondent to recall and construct 10 self-defining memories. Next, the respondent is asked to recall his/her first self-defining memory and to use a rating scale ranging from 0 (Not At All) to 6 (Extremely) to indicate how he/she felt in recalling and thinking about his/her memory. This rating scale includes rating the following emotions: happy, sad, angry, fearful, surprised, ashamed, disgusted, guilty, interested, embarrassed, contemptful, and proud. It also asks the respondent to indicate the vividness and importance of the memory and the approximate number of years since the memory took place.

17.3 CONCLUSION

Personality research methods of inquiry and personality tests used to assess personality can be used to advance the science of cybersecurity by applying and generating personality methods to better understand, protect, and defend information and systems in cybersecurity situations. A byproduct of this behavioral cybersecurity research can be advancing the knowledge base not only in the emerging field of behavioral cybersecurity, but also in the field of personality psychology. Also, with this knowledge about personality research methods, cybersecurity scholars and practitioners can explore how these methods can be applied to design future studies whose findings can be used to better describe, explain, and predict behavior of both the cyberattacker and defender in cybersecurity cases and scenarios.

REFERENCES

Alexander, I. 1990. *Personology: Method and Content in Personality Assessment and Psychobiography.* Durham, NC: Duke University Press.

Atkinson, J. W. 1958. Thematic apperceptive measurement of motives with the context of a theory of motivation. In J. W. Atkinson (Ed.), *Motives in Fantasy, Action and Society* (pp. 596–616). Princeton, NJ: Van Nostrad.

Atwood, G. and Stolorow, R. 1993. *Faces in a Cloud: Intersubjectivity in Personality Theory,* 2nd ed. Northvale, NJ: Jason Aronson.

Bess, T. L. and Harvey, R. J. 2002. Bimodal score distributions and the Myers-Briggs type indicator: Fact or artifact? *Journal of Personality Assessment*, 78(1), 176–186.

Briggs, K. and Myers, I. 1976. *The Myers-Briggs Type Indicator.* Palo Alto, CA: Consulting Psychologists Press.

Conner, T. S., Tennen, H., Fleeson, W., and Barrett, L. F. 2009. Experience sampling methods: A modern idiographic approach to personality research. *Social and Personality Psychology Compass*, 3(3), 292–313.

Crotty, M. 1998. *The Foundations of Social Research Meaning and Perspective in the Research Process*. London: SAGE Publications Inc.

Elms, A. C. 1986. From House to Haig: Private life and public style in American foreign policy advisers. *Journal of Social Issues*, 42(2), 33–53.

Elms, A. C. 2007. Psychobiography and case study methods. In B. Robins, C. Fraley and R. Krueger (Eds.), *Personality Research Methods* (pp. 97–113). New York: Guilford Press.

Elms, A. C. and Heller, B. 2005. Twelve ways to say "lonesome": Assessing error and control in the music of Elvis Presley. In W. T. Schultz (Ed.), *Handbook of Psychobiography* (pp. 142–157). New York: Oxford University Press.

Emmons, R. A. 1986. Personal strivings: An approach to personality and subjective wellbeing. *Journal of Personality and Social Psychology*, 51, 1058–1068.

Emmons, R. A. 1989. The personal striving approach to personality. In L. A. Pervin (Ed.), *Goal Concepts in Personality and Social Psychology* (pp. 87–126). Hillsdale, NJ: Lawrence Erlbaum Associates.

Horner, M. S. and Fleming, J. 1977. Revised scoring manual for an empirically derived soring system for the motive to avoid success. Unpublished manuscript, Harvard University, Cambridge, MA.

Jackson, D. N., Paunonen, S. V., Fraboni, M., and Goffin, R. D. 1996. A five-factor versus six-factor model of personality structure. *Personality and Individual Differences*, 20, 33–45.

Jung, C. G. 1971. *Psychological Types*. Princeton, NJ: Princeton University Press. (Original work published 1923.)

Kluckhohn, C. and Murray, H. A. 1953. Personality formation: The determinants. In C. Kluckhohn, H. A. Murray and D. M. Schneider (Eds.), *Personality in Nature, Society, and Culture* (pp. 53–67). New York: Knopf.

Larson, R. and Csikszentmihalyi, M. 1983. The experience sampling method. *New Directions for Methodology of Social and Behavioral Science*, 15, 41–56.

Lester, B., Lester, D., Wong, W. W. M., Cappelletti, D., and Jimenez, R. A. 2006. Some personality correlates of using eBAY. *Psychological Reports*, 99(3), 762–762.

Little, B. R. 1983. Personal projects: A rationale and method for investigation. *Environment and Behavior*, 15(3), 273–309.

Mayer, J.D., Faber, M.A., and Xu, X. 2007. Seventy-five years of motivation measures (1930–2005): A descriptive analysis. *Motivation Emotion*, 31, 83–103.

McAdams, D. P. 1980. A thematic coding system for the intimacy motive. *Journal of Research in Personality*, 14, 413–432.

McAdams, D. P. 1984. Scoring manual for the intimacy motive. *Psychological Documents*, 14(2613), 7.

McAdams, D. P. 1985. *Power, Intimacy, and the Life Story: Personological Inquiries into Identity*. New York: Guilford Press.

McAdams, D. P. 1992. The intimacy motivation scoring system. In C.P. Smith (Ed.), *Motivation and Personality: Handbook of Thematic Content Analysis* (pp. 229–253). New York: Cambridge University Press.

McAdams, D. A. 1997. The Guided Autobiography Instrument. Foley Center for the Study of Lives. Retrieved from www.sesp.northwestern.edu/foley.

McAdams, D. P. 2006. The role of narrative in personality psychology today. *Narrative Inquiry*, 16, 11–18.

McAdams, D. P. 2007. On grandiosity in personality psychology. *American Psychologist*, 62, 60–61.

McAdams, D. P., Bauer, J. J., Sakaeda, A. R., Anyidoho, N. A., Machado, M. A., Magrino-Failla, K. et al. 2006. Continuity and change in the life story: A longitudinal study of autobiographical memories in emerging adulthood. *Journal of Personality*, 74, 1371–1400.

McAdams, D. P. and West, S. 1997. Personality psychology and the case study. *Journal of Personality*, 65, 757–783.

McClelland, D. 1958. Methods of measuring human motivation. In J. W. Atkinson (Ed.), *Motives in Fantasy, Action and Society* (pp. 518–552). Princeton, NJ: Van Nostrand.

McCrae, R. R. and Costa, P. T. 1989. Reinterpreting the Myers-Briggs Type Indicator from the perspective of the five-factor model of personality. *Journal of Personality*, 57(1), 17–40.

McCrae, R. R. and Costa, P. T., Jr. 1997. Personality trait structure as a human universal. *American Psychologist*, 52, 509–516.

McCrae, R. R. and Costa, P. T., Jr. 1999. A five-factor theory of personality. In L. Pervin and O. John (Eds.), *Handbook of Personality: Theory and Research* (pp. 139–153). New York: Guilford Press.

McKay, J. R. 1992. A coring system for affiliative trust-mistrust. In C.P. Smith (Ed.), *Motivation and Personality: Handbook of Thematic Content Analysis* (pp. 266–277). New York: Cambridge University Press.

Morgan, C. D. and Murray, H. A. 1935. A method for investigating fantasies. *Archives of Neurology and Psychiatry*, 32, 29–39.

Murray, H. A. 1938. *Explorations in Personality*. New York: Oxford University Press.

Myers I. B. and McCaulley M. H. 1985. *Manual: A Guide to the Development and Use of the Myers-Briggs Type Indicator.* Palo Alto, CA: Consulting Psychologists Press.

Nasby, W. and Read, N. W. 1997. The life voyage of a solo circumnavigator: Integrating theoretical and methodological perspectives [Special issue]. *Journal of Personality*, 65, 785–1068.

Pintrich, P. R., Smith, D. A., Garcia, T., and McKeachie, W. 1993. Reliability and predictive validity of the Motivated Strategies for Learning Questionnaire (MSLQ). *Educational and Psychological Measurement*, 53, 801–813.

Revelle, W. 2007. Experimental approaches to the study of personality. In B. Robins, C. Fraley and R. Krueger (Eds.), *Personality Research Methods* (pp. 37–61). New York: Guilford Press.

Revelle, W., Humphreys, M. S., Simon, L., and Gilliland, K. 1980. The interactive effect of personality, time of day, and caffeine: A test of the arousal model. *Journal of Experimental Psychology: General*, 109, 1–31.

Reynierse, J. H. 2000. The combination of preferences and the formation of MBTI types. *Journal of Psychological Type*, 52, 18–31.

Robins, R. W., Fraley, C., and Krueger, R. F. 2007. *Handbook of Research Methods in Personality Psychology*. New York: Guilford Press.

Runyan, W. 2005. How to critically evaluate alternative explanations of life events: The case of van Gogh's ear. In W. T. Schultz (Ed.), *Handbook of Psychobiography* (pp. 96–103). New York: Oxford University Press.

Schultz, W. T. 1999. The riddle that doesn't exist: Ludwig Wittgenstein's transmogrification of death. *Psychoanalytic Review*, 86, 281–303.

Singer, J. A. 2004. Narrative identity and meaning making across the adult lifespan: An introduction. *Journal of Personality*, 72(3), 437–459.

Singer, J. A. and Blagov, P. 2004. The integrative function of narrative processing: Autobiographical memory, self-defining memories, and the life story of identity. In D. R. Beike, J. M. Lampinen, and D. A. Behrend (Eds.), *Studies in Self and Identity. The Self and Memory* (pp. 117–138). New York: Psychology Press.

Smith, C. P. 1992. *Motivation and Personality: Handbook of Thematic Content Analysis.* Cambridge, England: Cambridge University Press.

Stewart, A. J., Franz, C., and Layton, L. 1988. The changing self: Using personal documents to study lives. *Journal of Personality*, 56(1), 41–74.

Vallerand, R. J., Pelletier, L. G., Blais, M. R., and Brière, N. M. 1992. The academic motivation scale: A measure of intrinsic, extrinsic, and motivation in education. *Educational and Psychological Measurement*, 52, 1003–1017.

Winston, C. E. 2011. Biography and life story research. In S. Lapan, M. Quartaroli, and F. Riemer (Eds.), *Qualitative Research: An Introduction to Designs and Methods* (pp. 106–136). New Jersey: Jossey-Bass.

Winston-Proctor, C. E. 2018. Toward a model for teaching and learning qualitative inquiry within a core content undergraduate psychology course: Personality psychology as a natural opportunity. *Qualitative Psychology*, 5(2), 243–262.

Winter, D. G. 1973. *The Power Motive*. New York: The Free Press.

Winter, D. G. 1992. A revised scoring system for the power motive. In C. P. Smith (Ed.), *Motivation and Personality: Handbook of Thematic Content Analysis* (pp. 313–315). New York: Cambridge University Press.

Woike, B., Gershkovich, I., Piorkowski, R., and Polo, M. 1999. The role of motives in the content and structure of autobiographical memory. *Journal of Personality and Social Psychology*, 76(4), 600–612.

Zirkel, S., Garcia, J., and Murphy, M. C. 2015. Experience-sampling research methods and their potential for education research. *Educational Researcher*, 1–10.

PROBLEMS

These problems are designed to make you think about the essential behavioral science concepts that have been discussed in this chapter. These problems could be used in a number of ways, including as individual thought exercises, group discussion questions, and/or to stimulate interest in new ways of thinking about behavioral cybersecurity and methods of assessment of the dimensions of human personality (e.g., personality traits, motivation, goals, and narrative identity) and understanding the whole person.

17.1 Which of the research designs and methods are most interesting to you to advance understanding of the person in cybersecurity cases and scenarios with which you are familiar?

17.2 Identify a recent case of a human hacker in which there is a lot of information about the person from multiple sources (e.g., news reports, case studies, etc.) or select a case within Chapter 4, "Recent Events," and use the case descriptions as a source. Using any one or a combination of the personality assessments, describe a behavioral cybersecurity research question that might be pursued to advance understanding of either the defender or human hacker.

18 Modular Arithmetic and Other Computational Methods

Critical to the understanding of almost all cryptographic methods is an understanding of the study of what are called "modular arithmetic systems."

Any cryptographic method must ensure that when we perform the encryption step, we must be able to arrive back at the original or plaintext message when we perform the decryption. As an immediate consequence, this implies that representing information in computer memory cannot be interpreted as in the computer context as floating-point numbers, or, in mathematical terminology, real or rational numbers. The problem that arises is that interpretation of the result of any computation in computer memory of real or rational numbers leads to an uncertainty in terms of the lowest-order bit or bits of that computation. Therefore, in general, if we were to manipulate real or rational numbers in a cryptographic computation, the fact that the lowest-order bits are indeterminate could mean that no true or exact inversion could be performed.

As a consequence, virtually all cryptosystems use the natural numbers or integers as the basis for computation. Indeed, since computations in the integers might apply an unlimited range, instead we almost always use a smaller, finite, and fully enumerable system derived from the integers that we generally refer to as "modular arithmetic."

18.1 Z_n OR ARITHMETIC MODULO n

A standard definition of the integers can be written as $\mathbf{Z} = \{ -\infty, \ldots, -2, -1, 0, 1, 2, \ldots, \infty \}$ with operations $+$, $\times$. Of course, this set $\mathbf{Z}$ is infinite, so we derive a finite variant of the integers that we refer to as the "integers modulo n," written $\mathbf{Z}_n$ and defined as

$$\mathbf{Z}_n = \{ 0, 1, 2, \ldots, n - 1 \};$$

and if a and b $\in \mathbf{Z}_n$, $a + b$ is defined as the remainder of $a + b$ when divided by n, and $a \times b$ is defined as the remainder of $a \times b$ when divided by n.

A set of elements with a binary operation (such as $\mathbf{Z}_n$ with $+$) forms a group $G = \{ a, b, \ldots \}$ if several conditions are satisfied:

1. CLOSURE: if a, b $\in$ G, so is $a + b$.
2. IDENTITY: there is a special element called the identity, i, such that $a + i = i + a = a$, for all a in G.

3. ASSOCIATIVITY: for all a, b, c ∈ G, a + (b + c) = (a + b) + c.
4. INVERSE: for all a, there exists some b (that we write b = −a) such that a + b = a + (−a) = i.

In the case of $\mathbf{Z}_n$ with the + operation, the identity is 0, and all four conditions are satisfied, so $\mathbf{Z}_n$ with addition forms a group.

Let's take an example, first $\mathbf{Z}_6$:

Addition

+	0	1	2	3	4	5
0	0	1	2	3	4	5
1	1	2	3	4	5	0
2	2	3	4	5	0	1
3	3	4	5	0	1	2
4	4	5	0	1	2	3
5	5	0	1	2	3	4

Multiplication

×	0	1	2	3	4	5
0	0	0	0	0	0	0
1	0	1	2	3	4	5
2	0	2	4	0	2	4
3	0	3	0	3	0	3
4	0	4	2	0	4	2
5	0	5	4	3	2	1

Now consider $\mathbf{Z}_7$:

Addition

+	0	1	2	3	4	5	6
0	0	1	2	3	4	5	6
1	1	2	3	4	5	6	0
2	2	3	4	5	6	0	1
3	3	4	5	6	0	1	2
4	4	5	6	0	1	2	3
5	5	6	0	1	2	3	4
6	6	0	1	2	3	4	5

Multiplication

×	0	1	2	3	4	5	6
0	0	0	0	0	0	0	0
1	0	1	2	3	4	5	6
2	0	2	4	6	1	3	5
3	0	3	6	2	5	1	4
4	0	4	1	5	2	6	3
5	0	5	3	1	6	4	2
6	0	6	5	4	3	2	1

What are the differences in the tables?

- An element (in any multiplication table) has a multiplicative inverse ⇔ if there is a 1 in the row corresponding to that element.
- Which elements have inverses in $\mathbf{Z}_6$?
- Which elements have inverses in $\mathbf{Z}_7$?
- What is the essential difference between 6 and 7?
- One is prime (all nonzero elements have inverses).
- The other is composite (certain nonzero elements do not have inverses).
- Indeed, the elements that do not have inverses are exactly those that have a common factor with the composite number.

In general, in a modular arithmetic system based on the number n, if n is a composite number, there will always be some pair of numbers a and b less than n whose product will be 0 in the multiplication table, and in this case neither a nor b will have an inverse in the multiplication operation in $\mathbf{Z}_n$; if, however, the modular system is based on a prime number—let's call it p—then every nonzero element in $\mathbf{Z}_p$ will have a multiplicative inverse.

The reason, of course, is that in the case that n is composite, if you take two factors of n, a and b, greater than 1, then $a \times b = n$, that is, $a \times b \equiv 0 \pmod{n}$; then you can see from the table that neither a nor b can have a multiplicative inverse.

The result of these observations is that in the case of a prime number p, every nonzero element in the system has an inverse under multiplication. Therefore, the $\mathbf{Z}_p$ system (omitting the zero in the case of multiplication) contains two group structures, one for addition and one for multiplication. If we can add one other condition (which is indeed satisfied for all n whether prime or composite) called the distributive law relating addition and multiplication, that is: $a \times (b + c) = a \times b + a \times c$ for all a, b, and c, then the overall modular system is considered a mathematical field.

Therefore, we can conclude that the modular systems $\mathbf{Z}_p$ are fields when p is prime, and $\mathbf{Z}_n$ are not fields when n is composite.

18.2 WARNING!!!

For persons who are not familiar with calculations in modular arithmetic, it is important not to fall into this common trap:

When in a modular arithmetic system, $\mathbf{Z}_n$, and you divide …
For example: in $\mathbf{Z}_7$, consider $4 \div 3$:

It is not: 1.33333…

There are no decimals in modular arithmetic systems in $\mathbf{Z}_7$, $4 \div 3 = 6$, because
$(4 \times 3^{-1}) \bmod 7 = (4 \times 5) \bmod 7 = 20 \bmod 7 = 6$

Prime numbers, again, are those with no proper divisors, for example, 2, 3, 5, … , 13, 17, … , 23, 29, …

For any natural number n, call $\phi(n)$ the Möbius ϕ-function. It counts how many numbers between 1 and n are relatively prime to n—that is, have no common factors greater than 1. Clearly, if the number n is a prime, it has no factors greater than 1, so $\phi(p) = (p - 1)$.

In general, for large numbers n, $\phi(n)$ is infeasible to compute. We know that if the number is a prime, p, then $\phi(p) = (p - 1)$. Also, if n is the product of only two primes p and q ($n = pq$), then $\phi(n) = (p - 1) \times (q - 1)$.

There is one extremely important result about the Möbius function that arises many times in cryptography. We'll state this without proof, but that can be found in any elementary college algebra book. For any n, if you construct the mod n system, and for any $a < n$ that is relatively prime to n (alternatively, the greatest common divisor or GCD of a and n is $GCD(a,n) = 1$). Then in this case, raising a to the $\phi(n)$ power gives:

$$a^{\phi(n)} \pmod{n} = 1.$$

This is sometimes called the "little Fermat theorem."

18.3 FINITE FIELDS

This system can also be thought of as the integers $\mathbf{Z}$, $\mathbf{Z}/(p)$, which means in this new system, we collapse all the values that have the same remainder mod p.

We saw that if p is a prime, the system $\mathbf{Z}_p$ has the special property that all nonzero elements have multiplicative inverses; that is, for any $a \neq 0$, there exists some b for which $a \times b \equiv 1 \pmod{p}$.

Such an algebraic system mod p with addition and multiplication is called a *field*. In fact, such a (finite) field can be defined for all prime numbers p.

We can go a little further with finite fields. We can define the system of all polynomials $\mathbf{Z}_p[x]$ in a single variable, then $\mathbf{Z}_p[x]/(q(x))$, where q(x) is an *irreducible polynomial* of degree n.

The addition and multiplication of polynomials is as usual, except that the coefficients of the polynomials in the system are always modulo p.

So, for example, in the modulo 13 system of polynomials $Z_{13}[x]$, if we have $p1 = (3x^2 + 4x + 2)$ and $p2 = (x^3 + 5x^2 + 10x + 5)$, then

$$\begin{aligned} p1 + p2 &= (3x^2 + 4x + 2) + (x^3 + 5x^2 + 10x + 5) = (x^3 + 8x^2 + 14x + 7) \\ &= (x^3 + 8x^2 + x + 7) \pmod{13} \end{aligned}$$

$$\begin{aligned} p1 \times p2 &= (3x^2 + 4x + 2) \times (x^3 + 5x^2 + 10x + 5) \\ &= 3x^5 + (15+4)x^4 + (1+20+30)x^3 + (15+40+10)x^2 + (20+20)x + 10 \\ &= 3x^5 + 19x^4 + 51x^3 + 65x^2 + 40x + 10 \\ &= 3x^5 + 6x^4 + 12x^3 + 0x^2 + x + 10 \pmod{13} \end{aligned}$$

Irreducible polynomials q(x) are like prime numbers—they cannot be factored (beyond factoring coefficients). By analogy, the system where we collapse polynomials with the same remainder mod q(x) also becomes a field, which we call GF(p,n), the Galois field. Again, p is the mod system for the coefficients, and n indicates the degree of the polynomials—once we divide by q(x), no term will remain with an exponent higher than (n − 1).

Of the GF(p,n), several of the form GF(2,n) are the key components of the current U.S. government standard for data encryption, known as the Advanced Encryption Standard (AES) or Rijndael.

18.4 THE MAIN RESULT CONCERNING GALOIS FIELDS

The theory of these systems was a result developed by Evariste Galois, in the early nineteenth century, stating that all algebraic *fields* with a finite number of elements can be described as a GF(p,n) (including the fields $Z_{p,}$ since they can be thought of as GF(p,1), that is, dividing by an irreducible polynomial of degree 1 [such as ax + b]) (Figure 18.1).

Furthermore, all of the possible choices for a Galois field of type GF(p,n) are equivalent, and their number of elements is p^n.

And a bit about Galois himself: he lived in the early nineteenth century in Paris (Bell 1937). He developed these very important results in algebra while a teenager. He was also a political radical and went to prison. Upon his release, his interest in a young woman led to a duel in which he was killed at age 21. This was really a setup, since the supposed boyfriend of the romantic interest was actually a plant, one of the best sharpshooters in the French army.

He didn't name Galois fields; they were named after him.

18.5 MATRIX ALGEBRA OR LINEAR ALGEBRA

A matrix or array is a set of numbers of some type (integers, rational, or real, for example) considered as a rectangular set with a certain number of rows or columns. We say that a matrix is of order m × n if it has m rows and n columns and consequently has m × n elements all together.

FIGURE 18.1 Evariste Galois.

Here is an example of a 4 × 3 matrix A with real number values. We usually write a matrix enclosing its values in square brackets.

$$A = \begin{bmatrix} 3.7 & 1.6 & -2.9 \\ 6.8 & -4.3 & 0.7 \\ 2.6 & 5.9 & -3.7 \\ -1.4 & 2.4 & 9.3 \end{bmatrix}$$

The usual compact notation for A is $[a_{ij}]$, where i and j enumerate the elements in the rows and columns, respectively.

Under certain conditions, the operations of the system of the matrix elements can be extended to an operation on the matrices themselves. First, regarding addition, two matrices can only be added if they have the same dimension, m × n. In such a case:

Example:

$$A = \begin{bmatrix} 6 & 1 & 7 \\ 2 & -4 & 2 \end{bmatrix} \quad B = \begin{bmatrix} 5 & -3 & 4 \\ 7 & 6 & -2 \end{bmatrix}$$

$$A+B=\begin{bmatrix}6+5 & 1-3 & 7+4\\ 2+7 & -4+6 & 2-2\end{bmatrix}=\begin{bmatrix}11 & -2 & 11\\ 9 & 2 & 0\end{bmatrix}$$

Matrices can also be multiplied. The conditions for being able to do this are if you have A (with m rows and n columns) and B (with n rows and p columns), then A and B can be multiplied to form a matrix C, with C having m rows and p columns.

$$A\times B=\begin{bmatrix}6 & 3\\ 4 & 2\\ -1 & 5\end{bmatrix}\times\begin{bmatrix}3 & 2 & 1 & 6\\ -1 & 4 & 3 & 2\end{bmatrix}$$

$$=\begin{bmatrix}6\times3+3\times(-1) & 6\times2+3\times4 & 6\times1+3\times3 & 6\times6+3\times2\\ 4\times3+2\times(-1) & 4\times2+2\times4 & 4\times1+2\times3 & 4\times6+2\times2\\ -1\times3+5\times(-1) & -1\times2+5\times4 & -1\times1+5\times3 & -1\times6+5\times2\end{bmatrix}$$

$$=\begin{bmatrix}15 & 24 & 15 & 42\\ 10 & 16 & 10 & 28\\ -8 & 18 & 14 & 4\end{bmatrix}$$

Why is this useful? On the one hand, it provides a useful and compact way of describing an array of values. But on the other hand, perhaps the best example of this notation is how we can use it to translate a system of linear equations into a single matrix equation.

In the special case where you have a set of n linear equations in n unknowns, replacing the set of equations by a single matrix equation AX = B leads to a method of solving the entire system by solving the one matrix equation. Necessarily, because there are n equations in n unknowns, the matrix A is of order n × n, also known as a square matrix (of order n).

Square matrices A have the property that in many cases an inverse A^{-1} can be found. When this is the case, then the inverse matrix can be applied to both sides of the matrix equation, yielding the solution for X. In other words, multiplying both sides by A^{-1} yields

$$A^{-1}AX=X=A^{-1}B$$

$$3x+2y-4z+5w=15$$
$$4x-3y+2z-7w=-24$$
$$5x+4y-z+6w=34$$
$$x-2y+6z+2w=23$$

$$\begin{bmatrix}3 & 2 & -4 & 5\\ 4 & -3 & 2 & -7\\ 5 & 4 & -1 & 6\\ 1 & -2 & 6 & 2\end{bmatrix}\begin{bmatrix}x\\ y\\ z\\ w\end{bmatrix}=\begin{bmatrix}15\\ -24\\ 34\\ 23\end{bmatrix}$$

There is a classical formula to find A^{-1}. It is due to LaPlace:

$$\det(A) = \sum_{j=1}^{n} (-1)^{i+j} a_{i,j} M_{i,j} = \sum_{i=1}^{n} (-1)^{i+j} a_{i,j} M_{i,j}$$

where A is a square matrix of order n, and $M_{i,j}$ is the minor, defined to be the determinant of the (n – 1) × (n – 1)-matrix that results from A by removing the ith row and the jth column. The expression $(-1)^{i+j}M_{i,j}$ is known as the cofactor. Once again, the recursive formula would show that the big-O of the determinant computation would be f(determinant of order n) = n × O(determinant of order (n – 1)), and therefore the overall computation would be O(n!).

Then, with the determinant, the inverse matrix is

$$A^{-1} = \frac{1}{\det(A)} A.$$

In the simple 3 × 3 case, there is a simpler method to find A^{-1}, as indicated by the following:

The first method we might have learned could have been Cramer's rule:

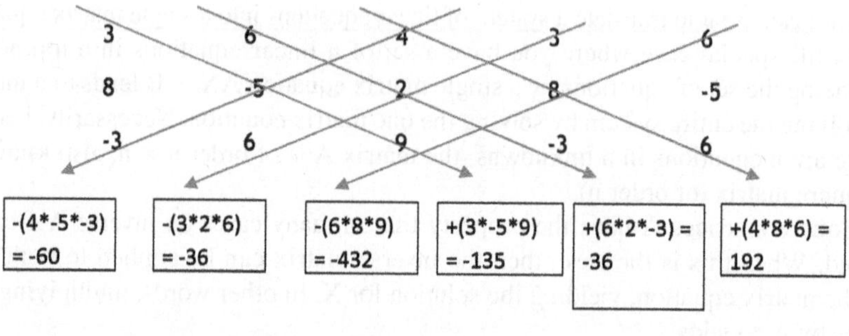

The arrows sloping down and to the right represent products of the three numbers preceded by a plus sign; the arrows sloping down and to the left are preceded by a minus sign. Then sum the individual products.

Therefore,

$$\det\begin{bmatrix} 3 & 6 & 4 \\ 8 & -5 & 2 \\ -3 & 6 & 9 \end{bmatrix} = -60 - 36 - 432 - 135 - 36 + 192 = 507.$$

However, this diagram does not extend beyond 3 × 3 matrices.

The more general approach to finding A^{-1} is called Gaussian elimination. It comes from writing the AX = B in what is called echelon form.

For example, consider the matrix equation

$$\begin{bmatrix} 9 & 3 & 4 \\ 4 & 3 & 4 \\ 1 & 1 & 1 \end{bmatrix} \begin{bmatrix} x_1 \\ x_2 \\ x_3 \end{bmatrix} = \begin{bmatrix} 7 \\ 8 \\ 3 \end{bmatrix}. \tag{18.1}$$

In augmented form, this becomes

$$\left[\begin{array}{ccc|c} 9 & 3 & 4 & 7 \\ 4 & 3 & 4 & 8 \\ 1 & 1 & 1 & 3 \end{array}\right] \begin{bmatrix} x_1 \\ x_2 \\ x_3 \end{bmatrix}. \tag{18.2}$$

Switching the first and third rows (without switching the elements in the right-hand column vector) gives

$$\left[\begin{array}{ccc|c} 1 & 1 & 1 & 3 \\ 4 & 3 & 4 & 8 \\ 9 & 3 & 4 & 7 \end{array}\right] \begin{bmatrix} x_1 \\ x_2 \\ x_3 \end{bmatrix}. \tag{18.3}$$

Subtracting 9 times the first row from the third row gives

$$\left[\begin{array}{ccc|c} 1 & 1 & 1 & 3 \\ 4 & 3 & 4 & 8 \\ 0 & -6 & -5 & -20 \end{array}\right] \begin{bmatrix} x_1 \\ x_2 \\ x_3 \end{bmatrix}. \tag{18.4}$$

Subtracting 4 times the first row from the second row gives

$$\left[\begin{array}{ccc|c} 1 & 1 & 1 & 3 \\ 0 & -1 & 0 & -4 \\ 0 & -6 & -5 & -20 \end{array}\right] \begin{bmatrix} x_1 \\ x_2 \\ x_3 \end{bmatrix}. \tag{18.5}$$

Finally, adding –6 times the second row to the third row gives

$$\left[\begin{array}{ccc|c} 1 & 1 & 1 & 3 \\ 0 & -1 & 0 & -4 \\ 0 & 0 & -5 & 4 \end{array}\right] \begin{bmatrix} x_1 \\ x_2 \\ x_3 \end{bmatrix}. \tag{18.6}$$

Restoring the transformed matrix equation gives

$$\begin{bmatrix} 1 & 1 & 1 \\ 0 & -1 & 0 \\ 0 & 0 & -5 \end{bmatrix} \begin{bmatrix} x_1 \\ x_2 \\ x_3 \end{bmatrix} = \begin{bmatrix} 3 \\ -4 \\ 4 \end{bmatrix}, \tag{18.7}$$

which can be solved immediately to give $x_3 = -4/5$, back-substituting to obtain $x_2 = 4$ (which actually follows trivially in this example), and then again back-substituting to find $x_1 = -1/5$.

REFERENCE

Bell, E. T. 1937. *Galois. Men of Mathematics 2.* New York: Simon & Schuster.

PROBLEMS

18.1 Solve the equations (or indicate if there is no solution):

$x^3 = 2 \pmod{15}$ Solution(s): ____________________

$x^2 + x + 1 = 0 \pmod{17}$ Solution(s): ____________________

$2^{10} \pmod{18}$ Solution(s): ____________________

$3^{1001} \pmod{40}$ Solution(s): ____________________

18.2 Solve $\sqrt{x} = 17 \pmod{29}$.
Solution(s): ____________________

18.3 Consider Z_{21} modular arithmetic, or mod 21 arithmetic. List all of the possibilities for an equation $a \times b \equiv 0 \pmod{21}$, where neither a nor b is 0 itself.

18.4 a. Create the multiplication table for $\mathbf{Z}_{15}$.
b. Solve the following for $\mathbf{Z}_{15}$:

7×8 ____________________

4^{-1} ____________________

$6^{-1} + (4 \times 7^{-1})$ ____________________

$8^2 \times 4^{-1}$ ____________________

$3 \div 8$ ____________________

$6 \div 5$ ____________________

18.5 Square elements in mod systems are interesting. Many nonzero elements in mod systems are not squares. Take as an example $\mathbf{Z}_7$. Note that 1, 2, and 4 are squares, because, for example, $6^2 = 1 \pmod 7$, $3^2 = 2 \pmod 7$, and $5^2 = 4 \pmod 7$. Also, you can show that 3, 5, and 6 are not squares.

18.6 Find all of the (nonzero) squares and nonsquares in mod 12 (or $\mathbf{Z}_{12}$).

Squares ____________________
Nonsquares ____________________

18.7 Display the calculations to find

GCD(8624,1837) = ______ and 1837^{-1} (mod 8624) = ____ Or check if it doesn't exist ______

GCD(89379,21577) = ____ and 21577^{-1} (mod 89379) = ____ Or check if it doesn't exist ______

GCD(438538,218655) = ___ and 218655^{-1} (mod 438538) = ____ Or check if it doesn't exist ____

18.8 Compute

a. $3 \div 4$ (mod 17) Solution:__________ Or doesn't exist______________
b. $18 \div 33$ (mod 121) Solution:__________ Or doesn't exist__________
c. $27 \div 16$ (mod 43) Solution:____________ Or doesn't exist__________
d. $12 \div 7$ (mod 15) Solution:_____________ Or doesn't exist__________

18.9 Calculate the Möbius function for n = 77. Find all the elements in $\mathbf{Z}_{77}$ that are not relatively prime to n.

18.10 For each of x = 33, 46, 49, 67, find the smallest exponent a for which $x^a \equiv 1 \pmod{n}$.

18.11 Multiply the following matrices (if this is possible):

$$\begin{bmatrix} 2 & 3 & 1 & 5 \\ -2 & 4 & 6 & 0 \end{bmatrix}$$

$$\begin{bmatrix} 1 & 0 & 5 \\ 2 & 7 & 1 \\ 6 & 6 & 2 \\ 3 & 8 & 3 \end{bmatrix}$$

18.12 Construct the Galois field GF(3,3). The elements will be all the polynomials with coefficients in $Z_3 = \{ 0, 1, 2 \}$, and polynomials of the form $ax^2 + bx + c$. The problem is to find an irreducible polynomial of degree 3, that is, $ax^3 + bx^2 + cx + d$. The search for an irreducible polynomial comes down to finding a third-degree polynomial under these definitions that does not factor into a second-degree poly times a first-degree poly considering that the coefficients are in Z_3. Once you find the irreducible polynomial, you can construct a multiplication table GF(3,3) since when you multiply lower-degree polynomials yielding a highest term of x^3 or higher, you can always reduce to a second-degree poly by dividing by the irreducible polynomial.

19 Modern Cryptography

In cryptography, the development of cryptographic techniques is inspired by what is called Kerckhoff's principle. This concept, established by Dutch cryptographer Auguste Kerckhoff in the nineteenth century, seems contradictory on first reading. It holds that almost all information about the cryptographic method should be revealed to the public, except for one component—the so-called *key* to the encryption. This long-established and reliable principle—also re-established for the modern electronic era by Claude Shannon of Bell Laboratories in the 1940s—is based on the concept that if we simply hid the cryptographic method, that is, the encryption or decryption algorithm, as soon as it was breached or obtained by bribery or other such means, the cryptographic system would be compromised.

On the other hand, the basic principle of steganography is that we embed a message in some enveloping vehicle. In olden times, before the electronic or computer era, it usually meant embedding one text inside another. One famous example is "Pershing." The following message was actually sent by a German Spy in WWII (Kahn, 1967):

- *Apparently neutral's protest is thoroughly discounted and ignored. Isman hard hit. Blockade issue affects pretext for embargo on byproducts, ejecting suets and vegetable oils.*

Taking the second letter in each word, the following message emerges:

Pershing sails from NY June 1.

Thus, the fundamental principle behind steganography is that we attempt to hide rather than encrypt secret messages so that the enemy will simply be fooled by thinking that an innocuous message really carries no hidden content. However, the weakness of this is that if an enemy maintains a suspicion that the message is not innocuous, it is usually far easier to find the hidden information and thus to break the steganograph.

Because we typically have less-than-perfect understanding of human behavior, it is very possible that there may be occasions on which a steganograph will operate successfully by fooling the opponent; on the other hand, a cryptographic method might fail if the opponent, knowing immediately that messages are scrambled or encoded, will know exactly what method is being used, and if the crypto method is weak enough, the attacker may be able to perform a successful decryption. One might ask, by what reason would the encryption be too weak? In general, the stronger an encryption method, the greater the cost to the encryptor in implementing it.

19.1 MODERN CRYPTOGRAPHIC TECHNIQUES

We have seen a number of historic approaches to cryptography in Chapter 10. However, once the digital age came upon us, the approach to cryptography changed dramatically. The understanding, once the communication of information became based on electronic transmission rather than on the printed word, was that what is referred to as the underlying alphabet of the system for encrypting and decrypting messages changed from the letters of a natural language (e.g., for the English language, { A, B, C, … , X, Y, Z }) to the language of the digital era, namely { 0, 1 }, the binary alphabet.

To underscore one difference between the two approaches, the structure of most human languages tends to differentiate in the usage of the underlying letters of such alphabets. For example, virtually any sufficiently large sample of text in English will demonstrate that the letter E will occur most often in the text, usually about 50% more often than the letter, usually T, that is the second most frequent in the text (Patterson, 1987).

There are however, some anomalous examples. The novel *A Void* by Georges Perec (1995) contains exactly zero uses of the letter E (regardless of upper- or lowercase) in its 290 pages. This novel was originally written in French with the same "no E" property. It was subsequently translated into English by Gilbert Adair.

Nonetheless, in our modern era, even though there are many cleverly designed cryptosystems, for the purposes of this book, we will focus on two in some detail. A third, which dominated modern cryptography for many years, called the Data Encryption Standard or DES, we will not discuss in detail but will try to indicate its place in the present time.

The United States government in the early 1970s came to realize that there was a need for the establishment of an official U.S. government standard for the encryption of data within U.S. government communications for civilian purposes. The military had already established other means, as indicated in the earlier chapter regarding one-time pads.

A collaboration developed between the National Security Agency, the government agency National Bureau of Standards (which has now been renamed the National Institute of Standards and Technology or NIST), and IBM to develop a proposal for a standard for data encryption. In 1975, the standard was published as the Data Encryption Standard. It used the principles of both transposition and substitution, and it acted on a plaintext message divided into 64-bit pieces, with an encryption key of 56 bits and 16 individual transformations of successive 64-bit pieces to result in encrypted 64-bit messages. One clever aspect of the DES was that the decryption step was essentially the use of the 16 transformations in reverse order, but using the same 56-bit key (NBS, 1977).

From the publication of the DES in 1975 until the late 1990s, not only was the DES the only standard approved by the U.S. government, but it was also the source of vast and heated differences in the crypto-community challenging the design and security of this standard, often referred to as the "Crypto Wars." The drama involved in this 25-year debate could be the source matter for a book by itself, but that needs to be left to another time. One important point is that the standard did not allow for variation in the usage of DES: the only approved method for use involved the division of the message into 64-bit pieces, and the key could only be 56 bits in length. For one thing, this meant that the size of the set of possible keys was limited to 2^{56} possibilities. Certainly at the time, the technology available to attempt all of the possible keys ($2^{56} \cong 7.2 \times 10^{16}$) was not feasible. However, by the early 1990s, an approach referred to as differential cryptology (Biham and Shamir, 1993) demonstrated a method that could analytically reduce the number of cases to test to usually around 2^{40}, which then made it feasible to essentially try every case, the so-called "brute force" approach.

These developments led to U.S. government initiating a competition to find a new national standard, which was adopted in early 2001 and called the Advanced Encryption Standard, which continues to be used to this day. There are certain similarities between the DES and the AES, and indeed the older and no longer standard DES can still be found in many encryption products, but DES is no longer considered secure by the government for the encryption community.

Consequently, we will describe the AES in some detail.

19.2 THE ADVANCED ENCRYPTION STANDARD

In 1997, NIST began a process to establish a new standard. In a drastic reversal from the approach used in the development of the DES, a public call was issued throughout the world for the development of a new standard and inviting the submission of algorithms subject to some initial design criteria. Initially, 21 algorithms were submitted to NIST in the competition, and 15 were selected for further review. It is interesting to note that 10 of the original 15 were not from the United States but from any of nine different countries.

After a considerable period of public review, the 15 candidates were reduced to 5. These were called MARS, RC6, Rijndael, Serpent, and Twofish. At the time, there was a good deal of speculation that the United States government would never accept a national encryption standard that was not a U.S. product, and among the five finalists, neither Rijndael nor Serpent were American.

The final selection, after considerable international review by the cryptographic community, was Rijndael (pronounced rain-doll), which was a submission by two Belgians, Vincent Rijmen and Joan Daemen (the name of the cryptosystem was a fusion of their last names). They were affiliated, respectively, with the Catholic University of Louvain and the Proton Corporation.

Upon its adoption—shortly after the "9/11" attack, it was renamed the Advanced Encryption Standard, considered a U.S. government standard with a slightly restricted set of the configurations in the design of Rijndael.

Although there are many papers, books, articles about Rijndael/AES, the most thorough reference is *The Design of Rijndael* (Daemen and Rijmen, 2002). The official government Advanced Encryption Standard is found at NIST (2001).

AES is based on a design principle known as a substitution–permutation network, and is efficient in both software and hardware. AES is a variant of Rijndael that has a fixed block size of 128 bits and a key size of 128, 192, or 256 bits. By contrast, Rijndael *per se* is specified with block and key sizes that may be any multiple of 32 bits, with a minimum of 128 and a maximum of 256 bits.

AES operates on a 4 × 4 column-major order array of bytes, termed the *state*. Most AES calculations are done in a particular finite field.

The key size used for an AES cipher specifies the number of transformation rounds that convert the input, called the plaintext, into the final output, called the ciphertext. The number of rounds is as follows:

- 10 rounds for 128-bit keys
- 12 rounds for 192-bit keys
- 14 rounds for 256-bit keys

In Rijndael, all of the mathematics can be done in a system called a Galois field, as we have described above. However, it is not essential in the use of Rijndael to do computations in the Galois field, but simply to use the results of such computations by looking up into appropriate tables.

We begin with the initial text to be encrypted, broken into an appropriate number of bytes, in our case 16 bytes. This text, and each time it is transformed, will be called the State. We will normally represent the bytes throughout the algorithm as hexadecimal symbols, { 0, 1, 2, 3, 4, 5, 6, 7, 8, 9, a, b, c, d, e, f }, where the six letter representations correspond to the decimal numbers 10, … , 15. A hex or hexadecimal number (base 16) is represented by 4 bits—corresponding to the above {0000, 0001, 0010, 0011, 0100, 0101, 0110, 0111, 1000, 1001, 1010, 1011, 1100, 1101, 1110, 1111}. Since a byte is 8 bits, any byte can be represented by the 8 bits, and each half of the byte by 4 bits, or by two hexadecimal numbers. Thus, for example, the 8-bit representation of a certain byte may be 01011011, or 0101 1011 (thus two hex numbers), or 5b (hex). Also, 5b (hex) expressed as an integer is (5 × 16) + b = (5 × 16) + 11 = 91 (decimal).

A critical step in dealing with these hex numbers is to determine their logical XOR (⊕) bit by bit or byte by byte. We might have, for example

$$\begin{aligned} 91_{dec} \oplus 167_{dec} &= 5b_{hex} \oplus a7_{hex} = 0101\,0001 \oplus 1010\,0111 \\ &= 1111\,0110 = f6_{hex} = 15 \times 16 + 6 = 246_{dec} \end{aligned}$$

We include a complete table of hex digits with the XOR or ⊕ operation. (See Table 19.1.)

The cipher key is also pictured as a rectangular array with four rows and columns of bytes.

In the case of our example, we will choose a text or State of 128 bytes, a key of similar size, and 10 rounds in the encryption. The pseudo-C code for a round will be:

```
Round(State, ExpandedKey[i])
{
   SubBytes(State);
   ShiftRows(State);
   MixColumns(State);
   AddRoundKey(State, ExpandedKey[i]);
}
```

Let's take these four steps in order.

TABLE 19.1
Hex Table under XOR or ⊕

⊕	0	1	2	3	4	5	6	7	8	9	A	B	C	D	E	F
0	0	1	2	3	4	5	6	7	8	9	A	B	C	D	E	F
1	1	0	3	2	5	4	7	6	9	8	B	A	D	C	F	E
2	2	3	0	1	6	7	4	5	A	B	8	9	E	F	C	D
3	3	2	1	0	7	6	5	4	B	A	9	8	F	E	D	C
4	4	5	6	7	0	1	2	3	C	D	E	F	8	9	A	B
5	5	4	7	6	1	0	3	2	D	C	F	E	9	8	B	A
6	6	7	4	5	2	3	0	1	E	F	C	D	A	B	8	9
7	7	6	5	4	3	2	1	0	F	E	D	C	B	A	9	8
8	8	9	A	B	C	D	E	F	0	1	2	3	4	5	6	7
9	9	8	B	A	D	C	F	E	1	0	3	2	5	4	7	6
A	A	B	8	9	E	F	C	D	2	3	0	1	6	7	4	5
B	B	A	9	8	F	E	D	C	3	2	1	0	7	6	5	4
C	C	D	E	F	8	9	A	B	4	5	6	7	0	1	2	3
D	D	C	F	E	9	8	B	A	5	4	7	6	1	0	3	2
E	E	F	C	D	A	B	8	9	6	7	4	5	2	3	0	1
F	F	E	D	C	B	A	9	8	7	6	5	4	3	2	1	0

19.2.1 SubBytes

SubBytes is a simple byte-by-byte substitution from two tables that can be found at the end of the chapter:

Just for an example, suppose you wish to find the result of SubBytes for the byte [1010 0111] = a7. Look in the S_{RD} table to find S_{RD}(a7) = 89 = [1000 1001]. Should you need to invert that result, look in the S_{RD}^{-1} table to find S_{RD}^{-1}(89) = a7.

19.2.2 ShiftRow

ShiftRow takes each row of the State and does a circular shift by 0, 1, 2, 3 positions as follows:

As opposed to SubBytes being a substitution, ShiftRow is a transposition. None of the byte values are changed, just the position of many of the bytes.

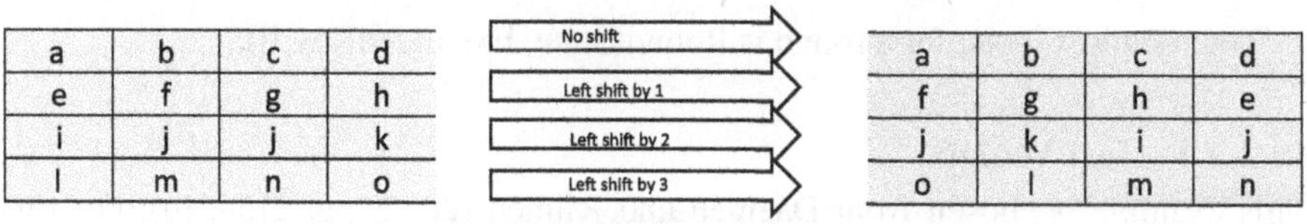

19.2.3 MixColumns

MixColumns introduces the main complexity in the overall algorithm. Viewed from the perspective of the underlying Galois fields, it is essentially a matrix multiplication.

But without going into the algebra of Galois fields, we can carry out this computation also as a form of matrix multiplication but where the "multiplication" of individual elements is essentially a logarithmic and antilogarithmic substitution, called "mul", and the "addition" along a row and column is the bitwise to bytewise exclusive-or operation.

19.2.4 AddRoundKey

The final step in an AES round is the selection of the portion of the key for the next round.

We are using AES in the mode with 128-bit (16-byte) key and test messages, and using the 10-round version.

Suppose the message or plaintext is as follows, in hex bytes:

```
32 43 f6 a8 88 5a 30 8d 31 31 98 a2 e0 37 07 34
```

and the key is:

```
2b 7e 15 16 28 ae d2 a6 ab f7 15 88 09 cf 4f 3c
```

The standard way of describing a Rijndael/AES encryption and decryption is with the terminology for each step being described as R[xx].yyyy, where the xx denotes a round, going from 00 to 10, and the yyyy represents the step, derived from the pseudocode:

R[00].input	only for round 00. This is the plaintext to be encrypted
R[00].k_sch	for the particular round (00 to 10), this is the key being used for this round
R[01].start	is simply the XOR of the plaintext and key
R[01].s_box	is the procedure called ByteSub using the single S-box
R[01].s_row	is the result of a procedure called ShiftRow
R[01].m_col	is the result of a procedure called MixColumn
R[01].k_sch	is the Key Schedule, or the generated key for the next round
R[02].start	is, again, the XOR of the result of round one and the key schedule generated at the end of round one

The pseudo-C code for a round is Round(State, ExpandedKey[i]).

19.2.4.1 Test Vectors

This example is chosen from Daemen and Rijmen (2002, pp. 215–216). For this example, we only compute one round.

This example assumes a 128-bit (or 16-byte) test message and cipher key.

Message or plaintext is (in hex bytes):

32 43 f6 a8 88 5a 30 8d 31 31 98 a2 e0 37 07 34

The key is:

2b 7e 15 16 28 ae d2 a6 ab f7 15 88 09 cf 4f 3c

Using the standard format for a trace of the encryption, we have:

```
R[00].input = 3 2 4 3 f 6 a 8 8 8 5 a 3 0 8 d 3 1 3 1 9 8 a 2 e 0 3 7 0 7 3 4
R[00].k_sch = 2 b 7 e 1 5 1 6 2 8 a e d 2 a 6 a b f 7 1 5 8 8 0 9 c f 4 f 3 c
R[01].start = _ _ _ _ _ _ _ _ _ _ _ _ _ _ _ _ _ _ _ _ _ _ _ _ _ _ _ _ _ _ _ _
R[01].s_box = _ _ _ _ _ _ _ _ _ _ _ _ _ _ _ _ _ _ _ _ _ _ _ _ _ _ _ _ _ _ _ _
R[01].s_row = _ _ _ _ _ _ _ _ _ _ _ _ _ _ _ _ _ _ _ _ _ _ _ _ _ _ _ _ _ _ _ _
R[01].m_col = _ _ _ _ _ _ _ _ _ _ _ _ _ _ _ _ _ _ _ _ _ _ _ _ _ _ _ _ _ _ _ _
R[01].k_sch = _ _ _ _ _ _ _ _ _ _ _ _ _ _ _ _ _ _ _ _ _ _ _ _ _ _ _ _ _ _ _ _
R[02].start = _ _ _ _ _ _ _ _ _ _ _ _ _ _ _ _ _ _ _ _ _ _ _ _ _ _ _ _ _ _ _ _
```

Now compute R[01].start by computing the XOR of R[00].input with R[00].k_sch:

```
32 43 f6 a8 88 5a 30 8d 31 31 98 a2 e0 37 07 34 ⊕
2b 7e 15 16 28 ae d2 a6 ab f7 15 88 09 cf 4f 3c =

19 3d e3 be a0 f4 e2 2b 9a c6 8d 2a e9 f8 48 08
```

So:

```
R[00].input = 3 2 4 3 f 6 a 8 8 8 5 a 3 0 8 d 3 1 3 1 9 8 a 2 e 0 3 7 0 7 3 4
R[00].k_sch = 2 b 7 e 1 5 1 6 2 8 a e d 2 a 6 a b f 7 1 5 8 8 0 9 c f 4 f 3 c
R[01].start = 1 9 3 d e 3 b e a 0 f 4 e 2 2 b 9 a c 6 8 d 2 a e 9 f 8 4 8 0 8
R[01].s_box = _ _ _ _ _ _ _ _ _ _ _ _ _ _ _ _ _ _ _ _ _ _ _ _ _ _ _ _ _ _ _ _
R[01].s_row = _ _ _ _ _ _ _ _ _ _ _ _ _ _ _ _ _ _ _ _ _ _ _ _ _ _ _ _ _ _ _ _
R[01].m_col = _ _ _ _ _ _ _ _ _ _ _ _ _ _ _ _ _ _ _ _ _ _ _ _ _ _ _ _ _ _ _ _
R[01].k_sch = _ _ _ _ _ _ _ _ _ _ _ _ _ _ _ _ _ _ _ _ _ _ _ _ _ _ _ _ _ _ _ _
R[02].start = _ _ _ _ _ _ _ _ _ _ _ _ _ _ _ _ _ _ _ _ _ _ _ _ _ _ _ _ _ _ _ _
```

19.2.4.2 Computing R[01].s_box

This is the SubBytes or S-box step. Note the S-box table, S_{RD}, can be found at the end of the chapter, as well as its inverse, S_{RD}^{-1} (Table 19.2).

The operation is simply to look up the S-box value for each byte in R[01].start. For example, the first hex pair of R[01].start is 19. In the S-box, the element at row 1, column 9 is d4.

Thus, the R[01].s_box is:

d4 27 11 ae e0 bf 98 f1 b8 b4 5d e5 1e 41 52 30

TABLE 19.2
SubByte S_{RD} and S_{RD}^{-1} Tables

S_{RD}

		y															
		0	**1**	**2**	**3**	**4**	**5**	**6**	**7**	**8**	**9**	**a**	**b**	**c**	**d**	**e**	**f**
	0	52	09	6a	d5	30	36	a5	38	bf	40	a3	9e	81	f3	d7	fb
	1	7c	e3	39	82	9b	2f	ff	87	34	8e	43	44	c4	de	e9	cb
	2	54	7b	94	32	a6	c2	23	3d	ee	4c	95	0b	42	fa	c3	4e
	3	08	2e	a1	66	28	d9	24	b2	76	5b	a2	49	6d	8b	d1	25
	4	72	f8	f6	64	86	68	98	16	d4	a4	5c	cc	5d	65	b6	92
	5	6c	70	48	50	fd	ed	b9	da	5e	15	46	57	a7	8d	9d	84
	6	90	d8	ab	00	8c	bc	d3	0a	f7	e4	58	05	b8	b3	45	06
	7	d0	2c	1e	8f	ca	3f	0f	02	c1	af	bd	03	01	13	8a	6b
x	8	3a	91	11	41	4f	67	dc	ea	97	f2	cf	ce	f0	b4	e6	73
	9	96	ac	74	22	e7	ad	35	85	e2	f9	37	e8	1c	75	df	6e
	A	47	f1	1a	71	1d	29	c5	89	6f	b7	62	0e	aa	18	be	1b
	B	fc	56	3e	4b	c6	d2	79	20	9a	db	c0	fe	78	cd	5a	f4
	C	1f	dd	a8	33	88	07	c7	31	b1	12	10	59	27	80	ec	5f
	D	60	51	7f	a9	19	b5	4a	0d	2d	e5	7a	9f	93	c9	9c	ef
	E	a0	e0	3b	4d	ae	2a	f5	b0	c8	eb	bb	3c	83	53	99	61
	F	17	2b	04	7e	ba	77	d6	26	e1	69	14	63	55	21	0c	7d

S_{RD}^{-1}

		y															
		0	**1**	**2**	**3**	**4**	**5**	**6**	**7**	**8**	**9**	**a**	**b**	**c**	**d**	**e**	**f**
	0	63	7c	77	7b	f2	6b	6f	c5	30	01	67	2b	fe	d7	ab	76
	1	ca	82	c9	7d	fa	59	47	f0	ad	d4	a2	af	9c	a4	72	c0
	2	b7	fd	93	26	36	3f	f7	cc	34	a5	e5	f1	71	d8	31	15
	3	04	c7	23	c3	18	96	05	9a	07	12	80	e2	eb	27	b2	75
	4	09	83	2c	1a	1b	6e	5a	a0	52	3b	d6	b3	29	e3	2f	84
	5	53	d1	00	ed	20	fc	b1	5b	6a	cb	be	39	4a	4c	58	cf
	6	d0	ef	aa	fb	43	4d	33	85	45	f9	02	7f	50	3c	9f	a8
	7	51	a3	40	8f	92	9d	38	f5	bc	b6	da	21	10	ff	f3	d2
x	8	cd	0c	13	ec	5f	97	44	17	c4	a7	7e	3d	64	5d	19	73
	9	60	81	4f	dc	22	2a	90	88	46	ee	b8	14	de	5e	0b	db
	a	e0	32	3a	0a	49	06	24	5c	c2	d3	ac	62	91	95	e4	79
	b	e7	c8	37	6d	8d	d5	4e	a9	6c	56	f4	ea	65	7a	ae	08
	c	ba	78	25	2e	1c	a6	b4	c6	e8	dd	74	1f	4b	bd	8b	8a
	d	70	3e	b5	66	48	03	f6	0e	61	35	57	b9	86	c1	1d	9e
	e	e1	f8	98	11	69	d9	8e	94	9b	1e	87	e9	ce	55	28	df
	f	8c	a1	89	0d	bf	e6	42	68	41	99	2d	0f	b0	54	bb	16

So now we have:

```
R[00].input = 3 2 4 3 f 6 a 8 8 8 5 a 3 0 8 d 3 1 3 1 9 8 a 2 e 0 3 7 0 7 3 4
R[00].k_sch = 2 b 7 e 1 5 1 6 2 8 a e d 2 a 6 a b f 7 1 5 8 8 0 9 c f 4 f 3 c
R[01].start = 1 9 3 d e 3 b e a 0 f 4 e 2 2 b 9 a c 6 8 d 2 a e 9 f 8 4 8 0 8
R[01].s_box = d 4 2 7 1 1 a e e 0 b f 9 8 f 1 b 8 b 4 5 d e 5 1 e 4 1 5 2 3 0
R[01].s_row = _ _ _ _ _ _ _ _ _ _ _ _ _ _ _ _ _ _ _ _ _ _ _ _ _ _ _ _ _ _ _ _
R[01].m_col = _ _ _ _ _ _ _ _ _ _ _ _ _ _ _ _ _ _ _ _ _ _ _ _ _ _ _ _ _ _ _ _
R[01].k_sch = _ _ _ _ _ _ _ _ _ _ _ _ _ _ _ _ _ _ _ _ _ _ _ _ _ _ _ _ _ _ _ _
R[02].start = _ _ _ _ _ _ _ _ _ _ _ _ _ _ _ _ _ _ _ _ _ _ _ _ _ _ _ _ _ _ _ _
```

19.2.4.3 Computing R[01].s_row

This is the Shift Rows step. Basically, one writes R[01].s_box into a 4-by-4 array, writing column-wise (i.e., fill the first column first, then the second column,…)

Writing column-wise				Now a circular left shift by 0, 1, 2, 3 places per line			
d4	e0	b8	le	d4	e0	b8	le
27	bf	b4	41	bf	b4	41	27
11	98	5d	52	5d	52	11	98
ae	fl	e5	30	30	ae	fl	e5

Shift left by x positions in row x (x = 0, 1, 2, 3). Now write this out in a single row to get R[01].s_row:

d4 bf 5d 30 e0 b4 52 ae b8 41 11 fl le 27 98 e5

In the previous notation,

```
R[00].input = 3 2 4 3 f 6 a 8 8 8 5 a 3 0 8 d 3 1 3 1 9 8 a 2 e 0 3 7 0 7 3 4
R[00].k_sch = 2 b 7 e 1 5 1 6 2 8 a e d 2 a 6 a b f 7 1 5 8 8 0 9 c f 4 f 3 c
R[01].start = 1 9 3 d e 3 b e a 0 f 4 e 2 2 b 9 a c 6 8 d 2 a e 9 f 8 4 8 0 8
R[01].s_box = d 4 2 7 1 1 a e e 0 b f 9 8 f 1 b 8 b 4 5 d e 5 1 e 4 1 5 2 3 0
R[01].s_row = d 4 b f 5 d 3 0 e 0 b 4 5 2 a e b 8 4 1 1 1 f 1 1 e 2 7 9 8 e 5
R[01].m_col = _ _ _ _ _ _ _ _ _ _ _ _ _ _ _ _ _ _ _ _ _ _ _ _ _ _ _ _ _ _ _ _
R[01].k_sch = _ _ _ _ _ _ _ _ _ _ _ _ _ _ _ _ _ _ _ _ _ _ _ _ _ _ _ _ _ _ _ _
R[02].start = _ _ _ _ _ _ _ _ _ _ _ _ _ _ _ _ _ _ _ _ _ _ _ _ _ _ _ _ _ _ _ _
```

19.2.4.4 Computing R[01]m_col

This is the Mix Columns step, undoubtedly the trickiest:

- This is actually a computation in the Galois field of polynomials over GF(2,8), that is, polynomials of degree 7, with binary coefficients.

- But let's not worry about that. It can also be expressed as a matrix product of a fixed matrix C. Suppose the State after ShiftRow is R[01].s_row written column-wise:

C	R[01].s_row
02 03 01 01	d4 e0 b8 1e
01 02 03 01	bf b4 41 27
01 01 02 03	5d 52 11 98
03 01 01 02	30 ae f1 e5

The result of this multiplication will be, as you know, another 4×4 matrix. Again, when we string out the column-wise version, we will get R[01].m_col.

However, these multiplications are in GF(2,8), or mod 256 arithmetic, and one can generate the "log tables" (see Rijmen and Daemen, 2001, pp. 221–222) to make the computation simpler.

Indeed, in the code is a brief function to do the multiplication (mul, p. 223). Essentially, mul is

$$\text{Alogtable}[(\text{Logtable}[a] + \text{Logtable}[b]) \text{ (mod) } 255]$$

19.2.4.5 Showing the calculation of the first byte

We will only compute the first byte of the matrix product, which is found by the usual method of the first row of the left-hand matrix by the first column of the right-hand matrix, thus:

02 03 01 01	d4 e0 b8 1e
01 02 03 01	bf b4 41 27
01 01 02 03	5d 52 11 98
03 01 01 02	30 ae f1 e5

Yielding for the first component (see Chapter 18):

$$02\ \text{d4} \oplus 03\ \text{bf} \oplus 01\ \text{5d} \oplus 01\ 30 = 02\ \text{d4} \oplus 03\ \text{bf} \oplus \text{5d} \oplus 30 \text{ (01 is the identity)}$$

Using the mul function for the first two terms (the right-hand side will be decimal numbers—see Table 19.3). That is, d4 in decimal is $13 \times 16 + 4 = 212$, and bf in decimal is $11 \times 16 + 15 = 191$:

$$\begin{aligned}\text{mul}(2,\text{d4}) &= \text{Alogtable}[\text{Logtable}[2] + \text{Logtable}[212]] \\ &= \text{Alogtable}[25 + 65] = \text{Alogtable}[90] = 179 = \text{b3(hex)};\end{aligned}$$

$$\begin{aligned}\text{mul}(3,\text{bf}) &= \text{Alogtable}[\text{Logtable}[3] + \text{Logtable}[191]] \\ &= \text{Alogtable}[1 + 157] = \text{Alogtable}[158] = 218 \\ &= \text{da}(\text{hex})\end{aligned}$$

TABLE 19.3
Logtable and Alogtable

	0	1	2	3	4	5	6	7	8	9	a	b	c	d	e	f
0	0	0	25	1	50	2	26	198	75	199	27	104	51	238	223	3
1	100	4	224	14	52	141	129	239	76	113	8	200	248	105	28	193
2	125	194	29	181	249	185	39	106	77	228	166	114	154	201	9	120
3	101	47	138	5	33	15	225	36	18	240	130	69	53	147	218	142
4	150	143	219	189	54	208	206	148	19	92	210	241	64	70	131	56
5	102	221	253	48	191	6	139	98	179	37	226	152	34	136	145	16
6	126	110	72	195	163	182	30	66	58	107	40	84	250	133	61	186
7	43	121	10	21	155	159	94	202	78	212	172	229	243	115	167	87
8	175	88	168	80	244	234	214	116	79	174	233	213	231	230	173	232
9	44	215	117	122	235	22	11	245	89	203	95	176	156	169	81	160
a	127	12	246	111	23	196	73	236	216	67	31	45	164	118	123	183
b	204	187	62	90	251	96	177	134	59	82	161	108	170	85	41	157
c	151	178	135	144	97	190	220	252	188	149	207	205	55	63	91	209
d	83	57	132	60	65	162	109	71	20	42	158	93	86	242	211	171
e	68	17	146	217	35	32	46	137	180	124	184	38	119	153	227	165
f	103	74	237	222	197	49	254	24	13	99	140	128	192	247	112	7

	0	1	2	3	4	5	6	7	8	9	a	b	c	d	e	f
0	1	3	5	15	17	51	85	255	26	46	114	150	161	248	19	53
1	95	225	56	72	216	115	149	164	247	2	6	10	30	34	102	170
2	229	52	92	228	55	89	235	38	106	190	217	112	144	171	230	49
3	83	245	4	12	20	60	68	204	79	209	104	184	211	110	178	205
4	76	212	103	169	224	59	77	215	98	166	241	8	24	40	120	136
5	131	158	185	208	107	189	220	127	129	152	179	206	73	219	118	154
6	181	196	87	249	16	48	80	240	11	29	39	105	187	214	97	163
7	254	25	43	125	135	146	173	236	47	113	147	174	233	32	96	160
8	251	22	58	78	210	109	183	194	93	231	50	86	250	21	63	65
9	195	94	226	61	71	201	64	192	91	237	44	116	156	191	218	117
a	159	186	213	100	172	239	42	126	130	157	188	223	122	142	137	128
b	155	182	193	88	232	35	101	175	234	37	111	177	200	67	197	84
c	252	31	33	99	165	244	7	9	27	45	119	153	176	203	70	202
d	69	207	74	222	121	139	134	145	168	227	62	66	198	81	243	14
e	18	54	90	238	41	123	141	140	143	138	133	148	167	242	13	23
f	57	75	221	124	132	151	162	253	28	36	108	180	199	82	246	1

Note: These two tables take a decimal input from 0 to 255, left to right. Express the input as 16a + b; then the appropriate value is in row a and column b.

Then, we need to compute b3 ⊕ da ⊕ 5d ⊕ 30

Or,

b	1011	3	0011
d	1101	a	1010
5	0101	d	1101
3	0011	0	0000

Or,

	0000		0100 = 04

After the full matrix multiplication, we have:

```
R[00].input = 3 2 4 3 f 6 a 8 8 8 5 a 3 0 8 d 3 1 3 1 9 8 a 2 e 0 3 7 0 7 3 4
R[00].k_sch = 2 b 7 e 1 5 1 6 2 8 a e d 2 a 6 a b f 7 1 5 8 8 0 9 c f 4 f 3 c
R[01].start = 1 9 3 d e 3 b e a 0 f 4 e 2 2 b 9 a c 6 8 d 2 a e 9 f 8 4 8 0 8
R[01].s_box = d 4 2 7 1 1 a e e 0 b f 9 8 f 1 b 8 b 4 5 d e 5 1 e 4 1 5 2 3 0
R[01].s_row = d 4 b f 5 d 3 0 e 0 b 4 5 2 a e b 8 4 1 1 1 f 1 1 e 2 7 9 8 e 5
R[01].m_col = 0 4 6 6 8 1 e 5 e 0 c b 1 9 9 a 4 8 f 8 d 3 7 a 2 8 0 6 2 6 4 c
R[01].k_sch = _ _ _ _ _ _ _ _ _ _ _ _ _ _ _ _ _ _ _ _ _ _ _ _ _ _ _ _ _ _ _ _
R[02].start = _ _ _ _ _ _ _ _ _ _ _ _ _ _ _ _ _ _ _ _ _ _ _ _ _ _ _ _ _ _ _ _
```

19.2.4.6 Last step—key schedule

In the key schedule, we use the previous key, XOR it with another part of the previous key, run through the S-box, with a possible counter added.

This time, I will only calculate the first word, or 4 bytes, of the key.

Take the first 4 bytes of the former key: 2b 7e 15 16

Left rotate once the last 4 bytes: 09 cf 4f 3c → cf 4f 3c 09

Run this last part through the S-box: SubByte(cf 4f 3c 09) = 8a 84 eb 01

XOR these, with a counter of 1 on the first byte:

$$2b \oplus 8a \oplus 01 \qquad 7e \oplus 84 \qquad 15 \oplus eb \qquad 16 \oplus 01 = \text{a0 fa fe 17}$$

The rest of the key schedule:

- On the previous slide, we determined the first 4 bytes of R[01].k_sch, namely: a0 fa fe 17
- The other 12 bytes are gotten by XORing 4 bytes at a time from the previous key, R[00].k_sch and the new key, as follows:
- `R[00].k_sch  2b 7e 15 16  28 ae d2 a6  ab f7 15 88  09 cf 4f 3c`
- `R[01].k_sch  a0 fa fe 17  88 54 2c b1  23 a3 39 39  2a 6c 76 05.`

```
R[00].input = 3 2 4 3 f 6 a 8 8 8 5 a 3 0 8 d 3 1 3 1 9 8 a 2 e 0 3 7 0 7 3 4
R[00].k_sch = 2 b 7 e 1 5 1 6 2 8 a e d 2 a 6 a b f 7 1 5 8 8 0 9 c f 4 f 3 c
R[01].start = 1 9 3 d e 3 b e a 0 f 4 e 2 2 b 9 a c 6 8 d 2 a e 9 f 8 4 8 0 8
R[01].s_box = d 4 2 7 1 1 a e e 0 b f 9 8 f 1 b 8 b 4 5 d e 5 1 e 4 1 5 2 3 0
R[01].s_row = d 4 b f 5 d 3 0 e 0 b 4 5 2 a e b 8 4 1 1 1 f 1 1 e 2 7 9 8 e 5
R[01].m_col = 0 4 6 6 8 1 e 5 e 0 c b 1 9 9 a 4 8 f 8 d 3 7 a 2 8 0 6 2 6 4 c
R[01].k_sch = a 0 f a f e 1 7 8 8 5 4 2 c b 1 2 3 a 3 3 9 3 9 2 a 6 c 7 6 0 5
R[02].start = a 4 9 c 7 f f 2 6 8 9 f 3 5 2 b 6 b 5 b e a 4 3 0 2 6 a 5 0 4 9
```

Getting to R[02].start consists of XORing R[01].m_col and R[01].k_sch.

Continue the same process for rounds 2, 3, … , 10.

19.3 THE KEY MANAGEMENT PROBLEM

However, despite the projections of security for the AES for some time into the future, it has one weakness it shares in common with almost every other encryption method described through the centuries. That weakness is the fact that the same key must be used for both encryption and decryption. The implication of this is that one presumes that normally encrypted messages need to travel in some fashion from the sender to the receiver, and therefore somehow that key must become known to both the sender and receiver.

Herein lies the problem that cannot be solved for AES or almost all of its predecessors. It is referred to as the key management problem, and the fact that that it must be shared by both sender and receiver has led all of the cryptographic methods with this requirement to be referred to generically as "symmetric encryption."

19.4 SYMMETRIC ENCRYPTION OR PUBLIC-KEY CRYPTOLOGY

Despite the pluses or minuses of AES, or any private key or symmetric method, one problem AES can never solve is the key management problem. Suppose we have a network with six users, each one of whom must have a separate key to communicate with each of the other five users. Thus, we will need in all for the six users (Figure 19.1):

$$\frac{6\times5}{2}=15 \text{ keys}$$

It is not unusual in these times to have a network of 1000 users, and thus we will need

$$\frac{1000\times999}{2}=499{,}500 \text{ keys.}$$

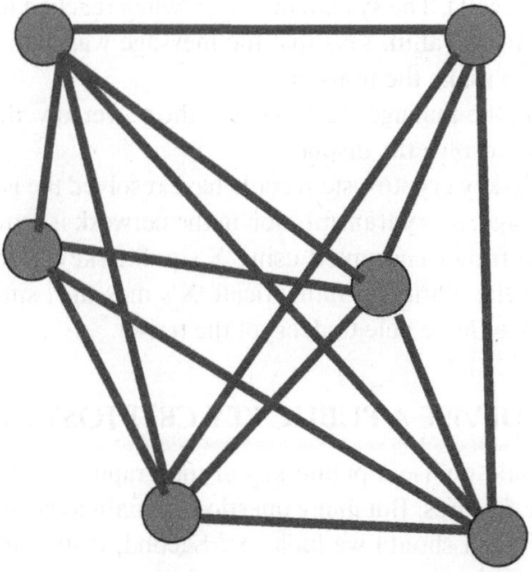

FIGURE 19.1 Complete graph with 6 nodes and 15 edges (6 nodes, (6 × 5)/2 = 15 edges).

And unfortunately, we cannot use this network to distribute the keys initially since until the keys are distributed, the network is insecure.

19.5 THE PUBLIC KEY CRYPTOGRAPHY MODEL FOR KEY MANAGEMENT

Now consider the case of the following approach. For each of the 1000 users, choose a key $k_i = (kp_i, ks_i)$, $i = 1, \ldots, 1000$. In a system-wide public directory, list all of the "public" keys kp_i, $i = 1, \ldots, 1000$. Then, to send a message m to user j, select the public key, kp_i, and apply the encryption transformation $c = T(kp_i, m)$.

Send the ciphertext, c.

Only user j has the rest of the key necessary to compute the decryption:

$$T((kp_i, ks_i), c) = m$$

Thus, rather than having to manage the secret distribution of $O(n^2)$ keys in a network of n users, only n keys are required, and they need not be distributed secretly.

Furthermore, the public-key concept could also be used for the authentication of messages in a way that a secret-key system could not address.

19.6 AUTHENTICATION

Consider a cryptosystem based on the traditional secret-key approach. Consider also that it is used for funds transfer in a banking network. One day the system manager receives a message from X. The manager decrypts the message using the secret key agreed upon by X and the manager. The message reads, "transfer $1,000,000 from my account to the system manager's account." The manager dutifully does so.

X complains to the authorities, saying that the message was a forgery, sent by the manager himself (herself). The system manager, when reached for comment by long-distance telephone from Tahiti, says that the message was authentic and that X had recanted his desire to make the transfer.

Since both X and the manager had to know the secret key, there is no way, using the cryptosystem, to resolve the dispute.

However, a public-key cryptosystem could have resolved the issue. Suppose that, in addition to the message, every transmission in the network is required to be "signed," that is, to contain a trailer encrypted using X's public key. Then, this requirement would carry with it the ability to authenticate X's message, since only X, knowing the rest of the key, would be able to decrypt the trailer.

19.7 CAN WE DEVISE A PUBLIC KEY CRYPTOSYSTEM

Therefore, if we could devise a public key cryptography (PKC), it would certainly have most desirable features. But many questions remain to be asked. First of all, can we devise a PKC? What should we look for? Second, if we can find one, will it be secure? Will it be efficient?

For now, we will consider only the general parameters of finding PKCs.

This approach, as described above, implies that the sender and receiver of encrypted information have two different parts of the key. Since the key k is broken into the so-called public key (k_p) and the secret key (k_s), and the public part is made available in the open for anyone to send a message to the creator of the key, and the secret key never leaves the creator, the creator and everyone else in the universe have different sets of information about the key, and therefore such a method is referred to as "asymmetric encryption."

Defining the problem in this way will certainly solve the key management problem; since only the public parts of everyone's key need be shared and can be placed in an open directory, and the private parts never have to leave the creator. But this is only a model, and does not describe how this asymmetric approach can be effectively carried out. This problem was solved by three cryptographers: Ron Rivest and Adi Shamir (from MIT), and Len Adelman from the University of Southern California, and was called the RSA Public Key Cryptosystem (Rivest et al., 1978).

19.8 THE RSA PUBLIC KEY CRYPTOSYSTEM

19.8.1 Factoring

From the earlier description, we need to find functions that are "one-way," that is, that enable an efficient computation sufficient for encryption, but whose inverses are cryptanalytically very difficult to find.

The example we will study involves the ease of multiplying numbers together combined with the difficulty of finding the original factors, given a product.

19.8.2 Who Was Pierre de Fermat?

He was a French mathematician of the seventeenth century (1601–1665). He purchased the offices of councillor at the parliament in Toulouse. This allowed him to change his name from "Pierre Fermat" to "Pierre de Fermat" (!).

He was a pioneer in geometry and number theory (Mahoney, 1994) (Figure 19.2).

19.9 FERMAT'S LAST THEOREM

Consider

$$3^2 + 4^2 = 5^2, (9 + 16 = 25)$$
$$5^2 + 12^2 = 13^2, (25 + 144 = 169)$$
$$7^2 + 24^2 + 25^2, (49 + 576 = 625)$$

Is there an example for $n > 2$ where $x^n + y^n = z^n$???

19.9.1 The 323-Year Marginalia

Late in his life, Fermat published a treatise on geometry, and before concluding, he wrote in a margin "I have discovered a truly remarkable proof which this margin is

FIGURE 19.2 Pierre de Fermat, 1601–1655, French mathematician.

too small to contain." Fermat's marginal comments led to many, many futile efforts to solve his "last theorem." It was finally solved in 1993 by Andrew Wiles. (The solution is that there are no values that would satisfy the equation for $n > 2$.)

19.10 THE LITTLE FERMAT THEOREM

As we saw in the chapter on modular arithmetic, the little Fermat theorem says that if I take any number, raise it to a power $\phi(n)$, and divide the result by n … I will get a remainder of 1. In mathematical notation,

$$a^{\phi(n)} \pmod{n} = 1$$

This is an essential result in the development of the following public-key cryptosystem.

19.11 THE RSA CRYPTOSYSTEM

About 30 years ago, three computer scientists, Ron Rivest, Adi Shamir, and Len Adleman, developed a public-key cryptosystem that they called the RSA cryptosystem (wonder why?). It is based entirely on the little Fermat theorem.

If Pierre de Fermat only could've gotten royalties, he could have bought all of Toulouse.

19.11.1 What Is the RSA Cryptosystem?

It is simple enough that the RSA Security Company has put it on a t-shirt (Figure 19.3).

- Take two prime numbers p and q (of 200 digits), and multiply $n = pq$.
- Find e and d such that their product gives a remainder of 1 when divided by $(p - 1)(q - 1)$.
- To encrypt, raise the message m to the power e (mod n).
- To decrypt, raise the cipher c to the power d (mod n).

19.11.2 Why You Should Be Skeptical …

1. Can we find prime numbers p, q of 200 digits? Well, not really. But we can use an algorithm, either the Solovay-Strassen or the Lehman-Peralta algorithms below, that will produce two numbers p and q that will be, except for once in every 2^{100} times. As the saying goes, good enough for government work.
2. Can we multiply them together? This could even be done by hand if you want to spend a whole afternoon doing this. By computer, less than a second.
3. Can we find an e such that $GCD(e, \varphi(n)) = 1$? In practice, picking e at random will work within a small number of tries (maybe a few hundred tries, but the computer won't care).
4. If we can find such an e, can we find d such that $e \times d \equiv 1 \pmod{\varphi(n)}$? Usually after a very few tries.
5. Can we realistically compute either m^e (mod n) or c^d (mod n)? See the fast exponentiation algorithm below.

19.12 PRIMALITY TESTING

19.12.1 If We Can't Factor Big Numbers …

How can we tell if they are prime?

The answer is, we can't … But we can choose a number p, which, if it passes a set of tests called "primality tests," we will be willing to accept as a prime, with a probability of $1/(2^{100})$ of guessing wrong.

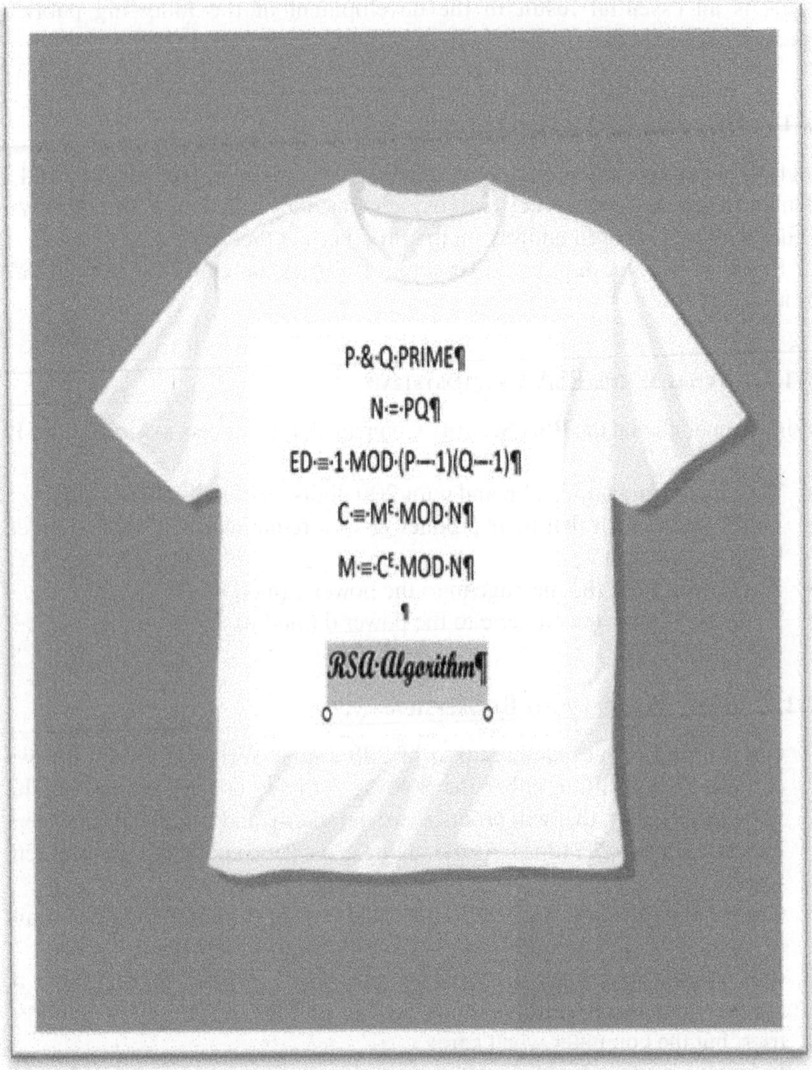

FIGURE 19.3 RSA Security t-shirt with RSA algorithm—facsimile.

Either the Solovay-Strassen or the Lehman-Peralta primality test (Solovay and Strassen, 1977) can be used.

To test p for primality, first choose 100 numbers at random <p, $e_1, \ldots, e_{100}$.

Compute GCD[e_i, p] for i = 1, … , 100.

If any GCD is >1, throw out p and start over!

For each e_i and p, compute a number called the Jacobi symbol. It is either 1 or −1. If the Jacobi symbol is 1, there is only a 50% chance that p is not prime and the Jacobi symbol is 1.

Thus, if all 100 Jacobi symbols are 1, there is only a $1/(2^{100})$ chance that p is not prime.

19.13 THE FAST EXPONENTIATION ALGORITHM

19.13.1 How Not to Compute $x^{16374927}$

```
Compute x × x × x × x × x × x × x × x × x × x × x × x × x ×
x × x × x × x × x × x × x × x × x × x × x × x × x × x × x ×
x × x × x × x × x × x × x × x × x × x × x × x × x × x × x ×
x × x × x × x × x × x × x × x × x × x × x × x × x × x × x ×
x × x × x × x × x × x × x × x × x × x × x × x ... × x
16,374,927 times
```

Maybe we can do this for 16 million … but of course, the former method will never complete if we're trying to compute

$x^{1728347619348723048762093847612471087462083476312O8}$

For example.

19.13.2 Fast Exponentiation for x^{14374}

First, convert 14374 to binary:

$$\begin{array}{r l} 14374 & \\ \underline{-8192} & (=2^{13}) \\ 6182 & \\ \underline{-4096} & (=2^{12}) \\ 2086 & \\ \underline{-2048} & (=211) \\ 38 & \end{array}$$

$$\begin{array}{r l} 38 & \\ \underline{-32} & (=2^{5}) \\ 6 & \\ \underline{-4} & (=2^{2}) \\ 2 & \\ \underline{-2} & (=2^{1}) \\ 0 & \end{array}$$

Therefore the binary is 11100000100110.
Now, $14374 = 11100000100110_2$

Call the bits $b_{13}b_{12}b_{11}...b_2b_1b_0$.
Ignore the high bit
To compute x^{14374},

```
set m = x
Do i = 12 downto 0
   m := m * m
   if bi = 1 then m = m * x
End do
m is the desired exponent.
```

Specifically:
m=x

$i = 12 : x \to x^2,$

$x^2 \times x = x^3$

$i = 11 : x^3 \to x^6,$

$x^6 \times x = x^7$

$i = 10 : x^7 \to x^{14}$

$i = 9 : x^{14} \to x^{28}$

$i = 8 : x^{28} \to x^{56}$

$i = 7 : x^{56} \to x^{112}$

$i = 6 : x^{112} \to x^{224}$

$i = 5 : x^{224} \to x^{448}$

$x^{448} \times x = x^{449}$

$i = 4 : x^{449} \to x^{898}$

$i = 3 : x^{898} \to x^{1796}$

$i = 2 : x^{1796} \to x^{3592}$

$x^{3592} \times x = x^{3593}$

$i = 1 : x^{3593} \to x^{7186}$

$x^{7186} \times x = x^{7187}$

$i = 0 : x^{7187} \to x^{14374}$

19.13.3 If Your Skepticism is Cured … Why Does It Work?

Aha! Little Fermat theorem:
$c^d = (m^e)^d$ (definition of decryption)
$\equiv m^{ed} \equiv$ (multiplication of exponents)
$\equiv m^{k\varphi(n)+1}$ (since $ed \equiv 1 \bmod \varphi(n)$)
$\equiv m^{k\varphi(n)} \times m^1$ (addition of exponents)
$\equiv 1 \times m = m$ (by the little Fermat theorem)
All computations mod n.

REFERENCES

Biham, E. and Shamir, A. 1993. Differential cryptanalysis of DES-like cryptosystems. *Journal of Cryptology*, 4(1), 3–72. Springer-Verlag.

Daemen, J. and Rijmen, V. 2002. *The Design of Rijndael.* New York: Springer-Verlag.

Kahn, D. 1967. *The Codebreakers.* New York: The Macmillan Company.

Mahoney, M. S. 1994. *The Mathematical Career of Pierre de Fermat, 1601–1665.* Princeton: Princeton University Press.

National Bureau of Standards (NBS). 1977. *Data Encryption Standard*, FIPS-Pub 46. Washington, DC.

National Institute for Standards and Technology (NIST). 2001. *Announcing the Advanced Encryption Standard (AES).* Federal Information Processing Standards Publication 197, November 26.

Patterson, W. 1987. *Mathematical Cryptology.* Totowa: Rowman and Littlefield.

Perec, G. 1995. *A Void* (English translation). London, England: The Harvill Press.

Rivest, R., Shamir, A., and Adleman, L. 1978. A method for obtaining digital signatures and public-key cryptosystems (PDF). *Communications of the ACM*, 21(2), 120–126. doi: 10.1145/359340.359342

Solovay, R. and Strassen, V. 1977. Fast Monte-Carlo tests for primality. *SIAM Journal on Computing*, 6(1), 84–85.

PROBLEMS

19.1 Find the SubBytes transformations for the bytes a5, 3f, 76, c9. Verify your results using the inverse SubBytes table.

19.2 Show the AES MixColumns transformation for

C	R[01].s_row
02 03 01 01	ef 7c 29 0b
01 02 03 01	c5 a6 33 d9
01 01 02 03	41 7a 99 e3
03 01 01 02	29 af 37 0d

19.3 Compute the indicated steps of an AES/Rijndael encryption when the message and key are given as below:

```
R[00].input = 0 1 0 2 0 3 0 4 0 1 0 2 0 3 0 4 0 1 0 2 0 3 0 4 0 1 0 2 0 3 0 4
R[00].k_sch = f 0 f 0 f 0 f 0 f 0 f 0 f 0 f 0 f 0 f 0 f 0 f 0 f 0 f 0 f 0 f 0
R[01].start = _ _ _ _ _ _ _ _ _ _ _ _ _ _ _ _ _ _ _ _ _ _ _ _ _ _ _ _ _ _ _ _
R[01].s_box = _ _ _ _ _ _ _ _ _ _ _ _ _ _ _ _ _ _ _ _ _ _ _ _ _ _ _ _ _ _ _ _
R[01].s_row = _ _ _ _ _ _ _ _ _ _ _ _ _ _ _ _ _ _ _ _ _ _ _ _ _ _ _ _ _ _ _ _
R[01].m_col = _ _ _ _ _ _ _ _ _ _ _ _ _ _ _ _ _ _ _ _ _ _ _ _ _ _ _ _ _ _ _ _
R[01].k_sch = _ _ _ _ _ _ _ _ _ _ _ _ _ _ _ _ _ _ _ _ _ _ _ _ _ _ _ _ _ _ _ _
R[02].start = _ _ _ _ _ _ _ _ _ _ _ _ _ _ _ _ _ _ _ _ _ _ _ _ _ _ _ _ _ _ _ _
```

19.4 We know that an RSA cryptosystem can be broken if the prime numbers p and q are small enough.

```
e = 3258563649422682316770352941747639 75263
n = 6617796424473520634885032706620161 40733
```

19.5 Convert 167845 decimal to binary.

19.6 Use the fast exponentiation algorithm to compute x^{167845}. Show all the steps.

20 Steganography

There have been many methods over time of steganographs (or steganograms) that conceal the existence of a message. Among these are invisible inks, microdots, character arrangement (other than the cryptographic methods of permutation and substitution), digital signatures, covert channels and spread-spectrum communications. As opposed to cryptography, steganography is the art of concealing the existence of information within innocuous carriers.

A message in ciphertext may arouse suspicion, while an *invisible* message will not. As a shorthand for the differences, cryptographic techniques "scramble" messages so if intercepted, the messages cannot be understood; steganography "camouflages" messages to hide their existence.

This one fact, in and of itself, suggests that the interface between cryptography and steganography needs to be explored within the context of behavioral science, since the approaches to creating and/or defending crypto or stego depend on human decisions based on their behavior.

David Kahn's *The Codebreakers* is a seminal history in this regard (1967), and the video (Stegano, 2011) provides a brief introduction.

20.1 A HISTORY OF STEGANOGRAPHY

One of the first documents describing steganography is from the *Histories* of Herodotus. In ancient Greece, text was written on wax-covered tablets. In one story, Demeratus wanted to notify Sparta that Xerxes intended to invade Greece. To avoid capture, he scraped the wax off the tablets and wrote a message on the underlying wood. He then covered the tablets with wax again. The tablets appeared to be blank and unused so they passed inspection by sentries without question.

Another ingenious method was to shave the head of a messenger and tattoo a message or image on the messengers head. After allowing his hair to grow, the message would be undetected until the head was shaved again. In modern parlance, this would be a pretty low-resolution methodology—perhaps a month to communicate a few bytes (Figure 20.1).

More common in more recent times, steganography has been implemented through the use of invisible inks. Such inks were used with much success as recently as World War II. Common sources for invisible inks are milk, vinegar, fruit juices, and urine, all of which darken when heated. These liquids all contain carbon compounds. When heated, the compounds break down and carbon is released, resulting in the chemical reaction between carbon and oxygen, that is, oxidation. The result of oxidation is a discoloration that permits the secret ink to become visible.

Null ciphers (unencrypted messages) were also used. The real message is "camouflaged" in an innocent-sounding message. Due to the "sound" of many open coded messages, the suspect communications were detected by mail filters. However

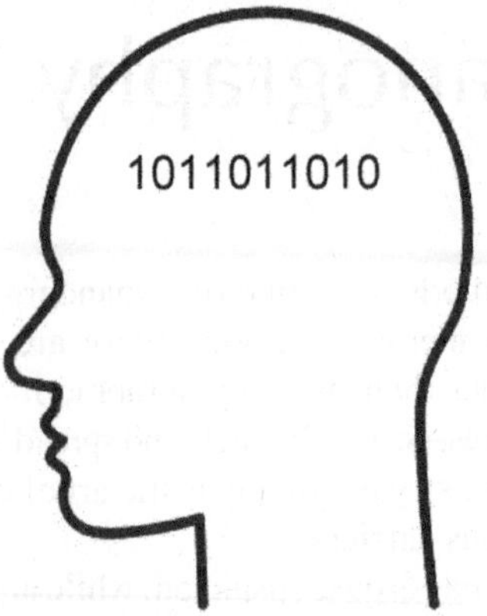

FIGURE 20.1 Very low-bandwidth steganography.

"innocent" messages were allowed to flow through. An example of a message containing such a null cipher is:

Fishing freshwater bends and saltwater coasts rewards anyone feeling stressed. Resourceful anglers usually find masterful leapers fun and admit swordfish rank overwhelming anyday.

By taking the third letter in each word in the quote, note the hidden message "SEND LAWYERS GUNS AND MONEY".

Suppose an obscure story appears on page 27 of the Sports section of the Oakland Tribune (CA):

However the baseball Athletics play ball, relievers cannot meet enviable needs for passing Houston's formidable array when winning under needy circumstances.

An alert reader might uncover the steganogram conveying Winston Churchill's famous exhortation:

*Ho**W**ever th**E** ba**S**eballAt**H**letics pl**A**y ball, re**L**ievers ca**N**not me**E**t en**V**iable ne**E**ds fo**R** pa**S**sing Ho**U**ston's fo**R**midable ar**R**ay wh**E**n wi**N**ning un**D**er ne**E**dy ci**R**cumstances.*

That is, taking the third letter of each word:

We shall never surrender!

The following message was actually sent from New York by a German spy in World War II:

Apparently neutral's protest is thoroughly discounted and ignored. Isman hard hit. Blockade issue affects pretext for embargo on byproducts, ejecting suets and vegetable oils.

Taking the second letter in each word the following message emerges:

APparently nEutral's pRotest iS tHoroughly dIscounted aNd iGnored. ISman hArd hIt. BLockade iSsue aFfects pRetext fOr eMbargo oN bYproducts, eJecting sUets aNd vEgetable oIls.

Or,

Pershing sails from NY June 1.

The Germans developed microdot technology, which FBI Director J. Edgar Hoover referred to as "the enemy's masterpiece of espionage" (Hoover, 1946).

Even the layout of a document can provide information about that document. Brassil et al. published several articles dealing with document identification and marking by modulating the position of lines and words (1995). Similar techniques can also be used to provide some other "covert" information, just as 0 and 1 are bits on a computer that can be concatenated in a string. Word-shifting can be used to help identify an original document. Though not applied as discussed in the series by Brassil et al., a similar method can be applied to display an entirely different message.

Take the following sentence (S0):

```
We explore new steganographic and cryptographic algorithms
and techniques throughout the world to produce wide variety
and security in the electronic web called the Internet.
```

and apply some word-shifting algorithm (this is sentence S1).

```
We  explore new steganographic and cryptographic algorithms
and techniques throughout  the  world to produce  wide variety
and security in the electronic  web called the Internet.
```

By overlapping S0 and S1, the following sentence is the result:

We **explore** new steganographic and cryptographic algorithms
and techniques throughout **the world** to produce **wide** variety
and security in the electronic **web** called the Internet.

This is achieved by expanding the space before *explore*, *the*, *wide*, and *web* by one point and condensing the space after *explore*, *world*, *wide* and *web* by one point in sentence S1. Independently, the sentences containing the shifted words appear harmless, but combining this with the original sentence produces a different message: *explore the world wide web.*

It should be noted that in the above example, we are making use of the Courier font, which has the property that the width of all characters are the same. This enables us to see more easily the extra blank space in front of certain words. This is normally not visible in many fonts, such as the Nemilov font used in this text.

20.2 TRANSMISSION ISSUES

Despite the long and interesting history of these various methods for hiding information, in practice they are declining in their importance.

One reason, although probably not the primary one, is the consequence of the use of a physical material to transmit information. Using invisible ink, to take one example, assumes that we have some medium on which this ink is deposited. Classically, this may be a letter with an innocuous message written on paper, and the invisible ink on top. Then the letter must reach its target, perhaps by postal service or courier. But who in these times would transmit such information on paper when electronic transmission is virtually instantaneous and capable of vastly larger messages or bandwidth?

Indeed, it may well be that the mere fact of transmission by a mail courier may arise suspicion, assuming the electronic means are readily available. Examples might include the threat a few years ago of anthrax contained in an envelope mailed to addresses in Washington, DC, including U.S. senators, and also the more recent example of bombs sent through the mail to former presidents Obama and Clinton.

However, a second and perhaps more important reason for the diminishing use of a physical medium such as paper for steganography is the challenge of embedding a message within text, such as the several examples given above—it is generally very difficult to encode!

The "fishing freshwater" example is a stego of 23 letters; the "apparently neutral's protest" hides 24 letters. It is an instructive exercise to try to construct a more detailed message of say, 100 letters, hidden inside text without the container being complete nonsense, thus raising suspicion.

20.3 IMAGE STEGANOGRAPHY

Consequently, with the current electronic age, the field of steganography has shifted to techniques of converting the secret message into a bitstring, then injecting the bitstring bit by bit not into text, but into some other file format such as an image file (e.g., JPEG, TIFF, BMP, or GIF) or a sound or movie file (MPEG, WAV, or AVI).

There are usually two type of files used when embedding data into an image. The innocent-looking image that will hold the hidden information is a "container." A "message" is the information to be hidden. A message may be plaintext, ciphertext, other images, or anything that can be embedded in the least significant bits (LSBs) of an image.

In this environment, for example, in an image file, the altering of a single bit in the image may be impossible to detect, certainly to the human eye, but also to an analysis of the file content byte by byte.

In order to do this, it is important to know something about standard image file formats (Eck, 2018; Wikipedia, 2018).

The essential component of a digital image, an image file, is the pixel (short form for "picture element"). The pixel represents the basic element of display on some medium such as a cathode ray tube, liquid crystal display, or even ink on printer paper.

If the medium of the display is capable of rendering color, the pixel values can indicate that color. The display related to an individual pixel is usually too small for the human eye to discern.

20.4 COMPRESSION

When file formats for digital images developed, a major concern was the size of the file representing the image. As a result, two strategies for reducing file size have been developed. These are called respectively "lossless compression" and "lossy compression" (Kurak and McHugh, 1992). Both methods save storage space but may present different results when the information is uncompressed.

Both approaches were devised in order to conserve storage by reducing the file size of the image. The techniques for lossless compression are of the nature that the complete original image can be reconstructed while yet reducing the size of the original image. Lossy compression techniques usually provide much more reduced file size; however, after compression, it is not possible to reconstruct the precise image pixel by pixel. As a consequence, care needs to be taken if a steganogram file size is then compressed, as a lossy compression approach may lead to a destruction of the hidden message.

20.5 IMAGE FILE FORMATS

The size of image files correlates positively with the number of pixels in the image and the color depth (bits per pixel). Images can be compressed in various ways, however. As indicated above, a compression algorithm stores either an exact representation or an approximation of the original image in a smaller number of bytes that can be expanded back to its uncompressed form with a corresponding decompression algorithm. Images with the same number of pixels and color depth can have very different compressed file sizes.

For example, a 640-by-480 pixel image with 24-bit color would occupy almost a megabyte of space:

$$640 \times 480 \times 24 = 7{,}372{,}800 \text{ bits} = 921{,}600 \text{ bytes} = 900\,\text{kB}$$

The most common image file formats are as follows.

JPEG (Joint Photographic Experts Group) is a lossy compression method. Nearly every digital camera can save images in the JPEG format, which supports 8-bit grayscale images and 24-bit RGB color images (8 bits each for red, green, and blue). JPEG lossy compression can result in a significant reduction of the file size. When not too great, the compression does not noticeably affect or detract from the image's quality, but JPEG files suffer generational degradation when repeatedly edited and saved.

TIFF (tagged image file format) format is a flexible format that normally saves 8 or 16 bits per RGB color for 24-bit and 48-bit totals, respectively, usually using either the TIFF or TIF filename extension. TIFFs can be lossy or lossless, depending on the technique chosen for storing the pixel data. Some offer relatively good lossless compression for bilevel (black and white) images. Some digital cameras can save images in TIFF format. TIFF image format is not widely supported by web browsers.

The BMP file format (Windows bitmap) handles graphic files within the Microsoft Windows OS. Typically, BMP files are uncompressed and therefore large and lossless; their advantage is their simple structure and wide acceptance in Windows programs.

FIGURE 20.2 Last James River ferry in operation. (Courtesy of Wayne Patterson.)

GIF (graphics interchange format) is in normal use limited to an 8-bit palette, or 256 colors (while 24-bit color depth is technically possible). GIF is most suitable for storing graphics with few colors, such as simple diagrams, shapes, logos, and cartoon-style images, as it uses lossless compression, which is more effective when large areas have a single color and less effective for photographic images. Because of GIF's simplicity and age, it achieved almost universal software support.

The following image is of the only remaining car ferry on the James River in Virginia (Figure 20.2).

The same image, when saved in the four image formats indicated above, requires considerably different space in kilobytes:

JPEG	712 KB
TIFF	6597 KB
BMP	18,433 KB
GIF	2868 KB

It is estimated that the human eye can distinguish perhaps as many as 3 million colors. If we use a fairly common color scheme or palette for an image to hide the steganograph, namely JPEG, there are 16.77 million possible RGB color combinations. Thus, we have many choices for altering the byte value associated with a pixel in order to conceal many bits of information and yet leave the image indistinguishable to the human eye.

Of course, if you have both the original image and the altered image, you don't have to rely on the human eye. You can use a "hex editor" (Hörz, 2018) to examine both the original image and the altered image byte by byte, and then it is a simple task to detect the differences.

It should also be noted: a JPEG file of a 4 × 6 image might be on the order of megabyte. A text to insert might be several kilobytes.

20.6 USING IMAGE STEGANOGRAPHY

Let's now describe how a dialogue could be carried on between two parties (eventually many parties) using steganography with secret text embedded in an appropriate JPEG image.

It is important to remember that there are two fundamental and conflicting considerations in the use of a cryptographic approach as opposed to a stenographic approach. Both of these conflicts deal fundamentally with an appreciation of the perceived behavior of the opponent or opponents.

In cryptography, we are very open in telling the opponent almost everything that we are doing. In other words, we are simply presenting a straightforward challenge and defying the opponent to put forth the effort to try to break the method. In steganography, we are trying to deceive the opponent into ignoring or perhaps applying less effort in order to intercept the message, recognizing, of course, that if the message is detected, no further effort is necessary because the opponent will be able to find the message.

Suppose our users are Angus and Barbara. Before they begin any messaging, they must come to some agreement on a container, and we suppose that it is a JPEG image I. Once this is accomplished, suppose Angus wants to send message m to Barbara. Angus will create the steganogram by inserting m into the container I; we will call the altered image or steganogram I'. Then Angus will send I' to Barbara.

What might be their strategies?

- I is in the clear
- I exists in some accessible database
- I is encrypted
- I is innocuous

Another aspect to consider in the use of images to contain a steganogram is whether it is likely to arouse suspicion for the transmission between the parties because it contains images rather than text. In other words, suppose that the normal dialogue between two parties when these messages are innocuous does not contain an image, and then suddenly a message is transmitted that does have an image. Is this a signal to the opponent that there is something special about the image and that might signal that in fact steganography is being used?

Returning to the strategies to initiate the dialogue between Angus and Barbara:

I is in the clear: If in the process of establishing the environment for steganography, we transmit the base image in the clear, an opponent can capture that image and then use it subsequently to detect the difference between a pure image and one that contains a steganogram, thus easily defeating our steganography.

I exists in some accessible database: Without transmitting the image in the clear, consider if it could be selected from some existing image database. Then we must assume that the opponent will have access to the same set of images and then could search that database in order to find the original and clear image to contrast the image containing the stego.

I is encrypted: Before we begin the transmission, the basic image or container could be encrypted and sent and thus both the sender and receiver will have the clear image for comparison and thus be able to detect the message within the stego. The disadvantage to this approach could be that the image file required is large, so the time for encryption and decryption of the image file could be significant. Of course, if the sender and receiver decide this is only necessary to do once in a relatively long period of time, this might not be a problem.

I is innocuous: As was indicated above, if it is rare for the sender and receiver to exchange image files, then the very existence of the transmission of an image might be a signal to the potential attacker that something is afoot. However, if the environment for transmission of messages between sender and receiver often contains some image, then it could be that the existence of an image itself might be sufficiently innocuous that it could pass muster and be ignored by the potential opponent.

The decision by Angus and Barbara as to how to establish what in cryptology we might call "key distribution" might be based on assumptions about the expected behavior of an attacker. Clearly, encrypting I would be the most secure method, but perhaps the most costly. The innocuous method could be an alternative, especially if we sense that the attacker is likely to be less diligent.

20.7 AN EXAMPLE

Example: We will use the easily available software tools HxD (Hörz, 2018) and QuickStego (Cybernescence, 2017).

We can present an example using two software tools, which any reader can download by himself or herself. These are both freeware products. One is a hexadecimal or hex editor, which allows the user to examine any file—therefore an image file byte by byte. There are many such editors available. One that we have used for this example is called HxD. The other software necessary for this example is again freeware, to insert a text into an image (thus creating a steganogram) where there are also numerous examples. The one we have chosen is called QuickStego (Figure 20.3).

It is often the case to use 256-color (or grayscale) images. These are the most common images found on the Internet in the form of GIF files. Each pixel is represented as a byte (8 bits). Many authors of steganography software stress the use of grayscale images (those with 256 shades of gray or better). The importance is

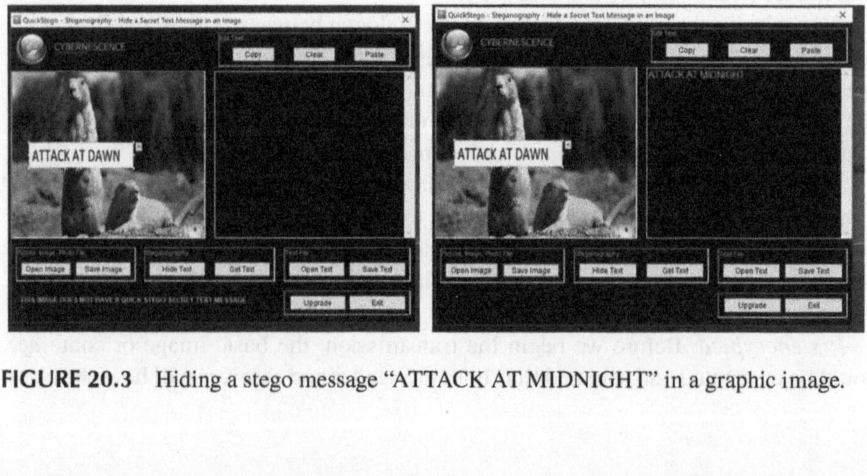

FIGURE 20.3 Hiding a stego message "ATTACK AT MIDNIGHT" in a graphic image.

not whether the image is grayscale, the importance is the degree to which the colors change between bit values.

Grayscale images are very good because the shades gradually change from byte to byte.

20.8 COMMENTS

Steganography has its place in security. It is not intended to replace cryptography but supplement it. As we will see, the existence of both cryptography and steganography leads to the possibility of hybrid techniques, which we will examine more deeply in the next chapter. Hiding a message with steganography methods reduces the chance of a message being detected. If that message is also encrypted, if discovered, it must also be cracked (yet another layer of protection). There are very many steganography applications. Steganography goes well beyond simply embedding text in an image. It does not only pertain to digital images but also to other media (files such as voice, other text and binaries; other media such as communication channels, and so on).

One area of considerable impact has been in "digital watermarking," which provides a form of copyright protection by inserting copyright information in a file steganographically.

REFERENCES

Brassil, J., Low, S., Maxemchuk, N., and O'Goram, L. 1995. Document Marking and Identification Using Both Line and Word Shifting. Infocom95, ftp://ftp.research.att.com/dist/brassil/1995/infocom95.ps.Z

Cybernescence. 2017. QuickStego. http://www.quickcrypto.com/free-steganography-software.html

Eck, D. J. 2018. Introduction to Computer Graphics. http://math.hws.edu/graphicsbook

Hoover, J. E. 1946. The enemy's masterpiece of espionage. *Reader's Digest*, 48, 1–6.

Hörz, M. 2018. HxD—Freeware Hex Editor and Disk Editor. https://mh-nexus.de/en/hxd/

Kahn, D. 1967. *The Codebreakers*. New York: The Macmillan Company.

Kurak C. and McHugh, J. 1992. A cautionary note on image downgrading. *IEEE Eighth Annual Computer Security Applications Conference*, pp. 153–159.

Steganonet. 2011. Evolution of Steganography. http://www.youtube.com/watch?v=osNWSGsFOvA

Wikipedia. 2018. Image File Formats. https://en.m.wikipedia.org/wiki/Image_file_formats

PROBLEMS

20.1 Create a steganogram that embeds the following 20-letter message in a larger text by using the same letter position in each word, either the first, second, third, or fourth. For example, if you were creating a steganogram for the message "HELLO," it might be "Have Everyone Leave Liquor Outside." Once you choose the letter position, all the words in your steganogram must use the same position, as in the example. (Ignore the blanks in the message below.)

FOURSCORE AND SEVEN YEARS AGO

20.2 Consider the principle of lossless vs lossy compression. Suppose you have an image wherein about 80% of the pixels represent the same color. How could you develop a coding system so that you would preserve all of the locations of the bytes of the same color, yet save a good deal of space in the rendering of the image?

20.3 Comment on the options available for Angus and Barbara in trying to establish a secure mechanism to exchange steganographs.

21 Using Cryptography and Steganography in Tandem or in Sequence

As we have seen, the cryptographic approach to secure messaging and the steganographic approach operate under two distinctly different and contradictory approaches. Cryptography is very open in telling any opponent even the technique or algorithm that is being used. In fact, in the RSA approach to public-key cryptology, any attacker can readily determine the size of the challenge in breaking the code because of the partial information involved in the public key. On the other hand, the steganographic approach attempts to appear completely normal, in the hopes that the attacker will be led to believe that there is no secret messaging involved and therefore will decide not to employ methods to try to determine if there is some secret to be revealed.

Even though these two approaches would seem to imply that the user must choose one or the other, there is a theory emerging that the two approaches of crypto and stego might be used in combination in various fashions.

Here is a very simple example. Suppose a person wishes to warn an ally of an impending attack. This person creates an image that clearly has the message "ATTACK AT DAWN." Then this person will send this to an ally. However, the message that will be clearly seen by anyone intercepting a message will warn of such an impending attack. However, what the sender has actually done, using, for example, the software QuickStego, is to embed the message "ATTACK AT MIDNIGHT."

The example is described in Figures 21.1 and 21.2.

Another more extensive example was presented in a recent master's thesis (Kittab, 2016) where the author, calling his approach Matryoshka Steganography (a matryoshka is one of the famous nested Russian dolls) used five levels of embedding in the hopes of deterring the person intercepting the message from burrowing five levels deep in order to find the true message. To quote from Kittab:

> We use the steganography to hide the significance of the message, and the probability of increasing the complexity of finding it in different media files with a different type of messages like text, image, audio, and video. This is a more secure way of communication because of using an unencrypted message container, which will never raise the flag of the importance of the message that is encoded into the unsuspicious message carrier.

This is a relatively new area of research and points out a more important aspect of how human behavior factors into all areas of cybersecurity.

The human factor involved in the use of cryptology arises when the defender, while telling the attacker everything but the key, relies upon the attacker's state of mind: since we assume that the attacker has as great a knowledge of cryptology as

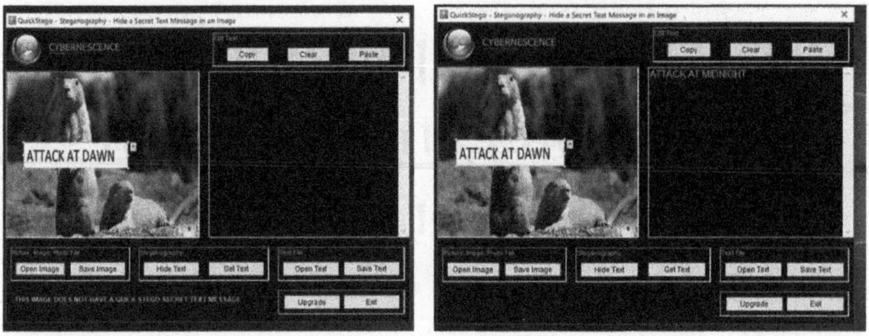

FIGURE 21.1 Hiding a stego message "ATTACK AT DAWN" in a graphics image.

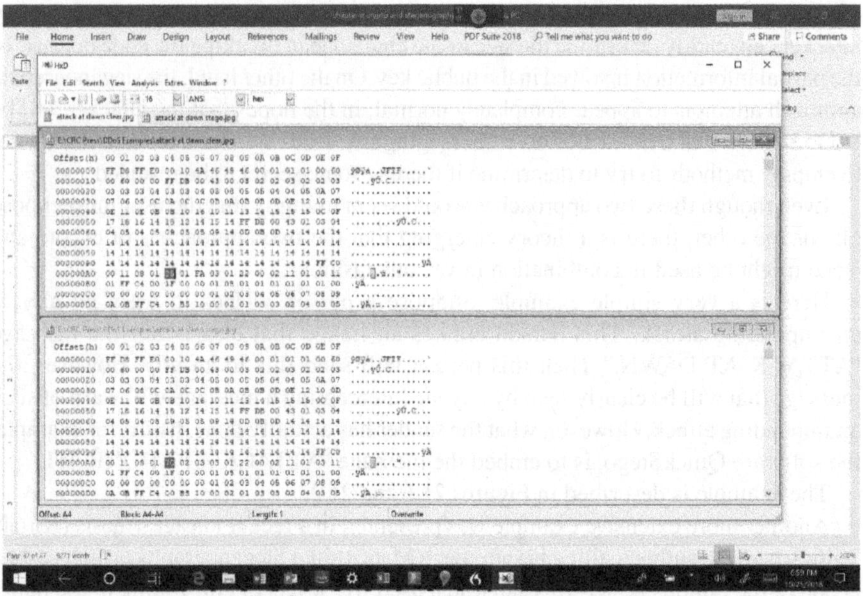

FIGURE 21.2 Viewing part of the clear and the stego in a hex editor.

the defender, the attacker can make a very determined calculation as to the cost of launching an attack—that is, trying to break the encryption—or to let it go by as not being worth the cost involved in deploying resources trying to break it.

On the other hand, the criterion for success of an embedded steganogram is a judgment on the attacker's behavior. If we were to assume that every attacker is unfailing in his or her determination to secure a message, we would surely be led to the cryptographic approach. However, in any environment, we are dealing with human beings, who operate on one or another behavioral model. If we were to learn more about who might be a potential attacker—for example, we might have certain known enemies—then we can develop a security strategy that is based on the human strengths and weaknesses that might result in an attack.

Then, together with a factor based on a calculated risk, the choice of security mechanism might be balanced between the crypto strategy, the stego strategy, or indeed a combination of both.

As indicated, this is a relatively new area of research, but it is clear that the analysis of the effectiveness of any such hybrid approach is not simply the calculation of the strength of the crypto or stego algorithms, but also the assessment of the human behavior of all the potential attackers.

REFERENCE

Kittab, W. M. 2016. *Matryoshka Steganography*. M.Sc. thesis, Howard University.

PROBLEMS

21.1 Create your own ATTACK AT DAWN, ATTACK AT MIDNIGHT example using QuickStego. Using a hex editor such as HxD, find all byte differences.

21.2 Manually embed a 30-byte message in a JPEG.

21.3 Use HxD to find all byte differences from your example in problem 2, from the position of the first difference.

22 A Metric to Assess Cyberattacks

With the rapid proliferation in recent times of vastly increased numbers of cyberattacks, especially those with extremely high profiles such as the events surrounding the 2016 Presidential election; the compromising of large corporate data sets such as with Sony, Equifax, and the U.S. government personnel office; and the recent spate of ransomware attacks, many individuals are now confused about what measures they can or should take in order to protect their information.

The various commercial solutions are often not helpful for the typical computer user. For example, some companies advertise protection against "all viruses," a claim that has been proven in research by Cohen (1987) to be theoretically impossible.

One feature of most attempts to protect computing environments is that they tends to be qualitative rather than quantitative in determining the level of protection that they may provide. This is undoubtedly because the challenge of determining the effort required of an attacker to penetrate one form of defense versus another is extremely difficult to compute, and thus there are no existing models to quantify that necessary effort.

For an average user, if such models were available, to be advised that a given cybersecurity package could only be defeated if the attacker was willing to spend \$10 million to deflect an attack, most users would not feel it would be necessary to have a level of defense of that order. An average user might be satisfied with a cyberdefense model that would deflect any attack that might cause the attacker to spend at least \$1 million in an attack effort.

This chapter does not attempt to solve the problem of an overall cyberdefense strategy that could provide such a metric, but we can give a model—in a very limited circumstance—where that level of protection can be determined very precisely.

It should be noted that the research developed in this chapter has had the invaluable assistance of one of the leading cybersecurity researchers in Africa, Professor Michael Ekonde Sone of the University of Buea in Cameroon (Sone and Patterson, 2017).

22.1 DEFINING A CYBERSECURITY METRIC

One of the main problems in trying to develop a definitive metric to describe a cybersecurity scenario that takes into account human decision-making is the difficulty in trying to measure the cost to either an attacker or defender of a specific approach to a cybersecurity event.

We can, however, describe one situation in which it is possible to describe very precisely the cost of engaging in a cybersecurity attack or defense. It should be noted that this example we will develop is somewhat unique in terms of the precision we can assign and therefore in the definitive nature of the assessment of the cost and benefit to each party in a cyberattack.

The case in point will be the use of the so-called Rivest-Shamir-Adleman public key cryptosystem, also known as the RSA. The RSA has been widely used for over 40 years, and we will provide only a brief introduction to this public key cryptosystem. However, in order to describe the metric, we will need to develop some applied number theory that will be very particular to this environment.

22.2 THE ATTACKER/DEFENDER SCENARIO

In terms of the overall approach to the problem, it is necessary to define a metric useful to both parties in a cyberattack environment, whom we will call the Attacker and the Defender. Often, of course, a more prevalent term for the Attacker is a "hacker," but we will use the other term instead. An attacker may be an individual, an automated "bot," or a team of intruders.

The classic challenge we will investigate will be formulated in this way: the Attacker is capable of intercepting various communications from the Defender to some other party, but the intercepted message is of little use because it has been transformed in some fashion using an encryption method.

With any such encrypted communication, which we also call ciphertext, we can define the cost of an attack (to the Attacker) in many forms. Assuming the attack takes place in an electronic environment, then the cost to decipher the communication will involve required use of CPU time to decipher, or the amount of memory storage for the decryption, or the Attacker's human time in carrying out the attack. It is well known in computer science that each of these costs can essentially be exchanged for one of the others, up to some constant factor. The same is true for the Defender.

For the sake of argument, let us assume that we will standardize the cost of an attack or defense by the computer time necessary to carry out the attack.

For one example, let us consider a simple encryption method, P, based on a random permutation of the letters of the alphabet. In this case, just to reduce the order of magnitude of the computation, we will reduce the alphabet to its most frequent 20 symbols, by eliminating the alphabet subset { J, K, Q, V, X, Z }. Assuming that the messages are in English, the particular choice of encryption could be any of the possible permutations of the modified Roman 20-letter alphabet, for example:

$$\{ \text{A B C D E F G H I L M N O P R S T U W Y} \}$$

$$\downarrow$$

$$\{ \text{T W F O I M A R S B U L H C Y D P G E N} \}$$

In other words, any A in the plaintext is replaced by a T, a B by a W, and so on. Expressed differently, any choice of the specific encryption transformation (called the key) also describes a permutation of a set of 20 objects.

Thus, there is a potential of the key being one of 20! choices, given that the cardinality of P, $|P| = 20!$. How might the Attacker try to intercept such an encrypted message? Let us make one further simplification, in that the Attacker correctly assumes that the Defender has chosen the encryption method P.

The simplest approach, therefore, is simply for the Attacker to test all of the potential choice of keys until the correct one is found. This is usually called "exhaustive search" or "brute force." By a simple probabilistic argument, the Attacker should only expect to have to try (20!)/2 keys before finding the correct one.

Thus, if our Attacker could establish the amount of time necessary to test one key as, for example, one microsecond, then for an exhaustive search attack to be successful would require:

$$\begin{aligned}(20!)/2 &= 1{,}216{,}451{,}004{,}088{,}320{,}000 \\ &= 1.216\times 10^{18}\,\mu s = 1.\,216\times 10^{12}\,s.\end{aligned}$$

In perhaps a more useful unit comparison, this would represent 38,573.4 years. If the Attacker could then estimate the cost of his or her computer time, memory, or human time as $.10 per key tested, then he or she can estimate the cost to break the encryption would be $1.22 $\times$ 10^{17}. A rational attacker would then conclude that if the value of the message to be decrypted was less than $$10^{17}$ (which would almost certainly be the case), the attack would not be worthwhile.

This could serve as our first approximation as a metric for the security of such an encrypted communication. Clearly both the Attacker and the Defender can make this computation, and if the Attacker feels that success in the attack is not worth the time or money spent, he or she will not attempt the attack.

Unfortunately, this is only a first approximation and makes an assumption that is fallacious. In this particular case, the Attacker has at his or her disposition other tools to break the message that are not dependent on the extreme cost of an exhaustive search attack.

For example, with this simple encryption method, the Attacker might logically conclude that the original (plaintext) message was in the English language, and therefore conduct what is known as a frequency analysis. Since we have assumed that each letter of the alphabet of the message is replaced by another, a frequency attack simply tallies the number of times each letter in the encrypted message is used. In any sample of English language text of a sufficient size, in almost every case, the letter "E" will occur almost 50% more often than any other letter of the alphabet. Hence, once the attacker has conducted the tally of all of the letters in the encrypted message (which would take only a few seconds even for very large body of text), he or she will have discovered the encrypt for the letter E. Also, in most cases, the second most frequent letter occurring will be the letter "T," so the Attacker can make that replacement as well.

Many newspapers, at least in North America, carry a daily "Cryptogram" which uses this particular encryption method, in which many people are challenged to solve normally only with paper and pencil. Certainly, capable Attackers trying to decipher a message as described above will always be more successful and in a much shorter time using frequency analysis rather than exhaustive search.

Consequently, one can conclude that in order to establish a metric that will be useful to both Attacker and Defender, it is necessary to understand all possible methods of decryption, and unfortunately, in many real-world cases, this is difficult if not impossible.

22.3 RIVEST-SHAMIR-ADLEMAN: AN INTERESTING EXAMPLE

There is one example where, despite many efforts over the past 40 years, it is possible to determine precisely the cost to an attacker to break a given encryption. This is the encryption method known as the RSA (or Rivest-Shamir-Adleman Public Key Cryptosystem) (Rivest et al., 1978).

The ingenuity in the definition of the RSA is that a very simple algebraic process can lead to the only known method of breaking the encryption through the factoring of a number n that is the product of two primes, p and q. It is also advisable that the p and q be numbers of the same number of decimal digits to avoid brute-force attack since they will be chosen in the middle of the dynamic range of number n.

22.4 CREATING THE RIVEST-SHAMIR-ADLEMAN PUBLIC-KEY CRYPTOSYSTEM

With any cryptosystem, the beginnings of the definition of the system constitute the phase that is generally called the key generation.

In the case of the RSA, the key generation begins with the selection of two prime numbers, say p and q, of 200 decimal digits, that is, $10^{200} < p, q < 10^{201}$, then their product $n = p \times q$.

In general, for composite numbers, the Möbius function $\phi(n)$, defined as the cardinality of the set of all divisors of n, is not possible to calculate for large integers. However, in the special case of a number being the product of two primes, $\phi(n)$ can be calculated as $\phi(n) = (p - 1) \times (q - 1)$.

The next step consists of finding two numbers, e (encryption key) and d (decryption key), which are relatively prime modulo $\phi(n)$, in other words, $e \times d \equiv 1 \pmod{\phi(n)}$.

This completes the key generation phase. With most encryption methods, described as symmetric encryption, both parties to the encryption process must possess the same key information. However, in recent years, methods known as public-key cryptosystems (or asymmetric cryptosystems) (Diffie and Hellman, 1976) have been developed, of which RSA is a prime example. In such cryptosystems, the "public" part of the key is made known to everyone, and the remainder of the key rests only with the creator and is never shared. In the case of the RSA cryptosystem we have just defined, suppose Alice has created the system; the e and n form the public key, and p, q and d form the private key, which Alice alone knows.

Once the public key values are made accessible to the public, anyone wishing to send a message M to Alice looks up her public key values of e and n, selects an appropriate number of bits m of the message M, interprets it as an integer less than n, and then encrypts by computing the cipher, c: $c \equiv m^e \pmod{n}$. The encryption proceeds by taking subsequent equal pieces of the bitstring M and performing the same operation. Once complete, all the pieces of ciphertext are strung together (call this C) and sent to Alice. Since Alice alone knows the secret parts of the overall key, in particular d, she selects each part of the cipher c and computes $m' \equiv c^d \pmod{n}$. Of course, in order for this to decrypt to find the original message, we need to demonstrate the simple algebraic step, which depends only on what is sometimes called the little Fermat theorem, namely that for $a \neq 0$ in a mod n system, $a^{\phi(n)} \equiv 1 \pmod{n}$.

$$
\begin{aligned}
c &\equiv m^e \pmod{n}; && \\
m' \equiv c^d &\equiv \left(m^e\right)^d \pmod{n}; && \text{(by definition);} \\
&\equiv m^{ed} \pmod{n}; && \text{(multiplicative law of exponents);} \\
&\equiv m^{k\phi(n)+1} \pmod{n}; && \text{(because e and d are inverses mod } \phi(n)\text{);} \\
&\equiv m^{k\phi(n)} \times m^1 \pmod{n}; && \text{(additive law of exponents);} \\
&\equiv 1 \times m = m \pmod{n}; && \text{(little Fermat theorem).}
\end{aligned}
$$

There are numerous questions that need to be addressed in both the key generation and in the operation of the RSA, but at least this shows that the decryption followed by the encryption returns to the original value.

The other questions involve the difficulty of finding prime numbers p and q of sufficient size, although this is not a significant challenge even if the primes were to be of several hundreds or even thousands of digits. In addition, finding the inverses e and d (mod $\phi(n)$) and the result of raising a several-hundred-digit number to an exponent that is also several hundred digits both need to be shown to be feasible of being computed in reasonable time. It is certainly possible to demonstrate this, as can be found in numerous references (Patterson, 2016).

Also, fortunately for this discussion, the RSA cryptosystem is highly scalable in that it can begin with an arbitrary number of digits for the choice of primes p and q (although it is advisable that p and q be roughly of the same size). Consequently, the designer of a crypto defense can choose these parameters and therefore parameterize the cost to a potential attacker based on the number of digits of p and q. Furthermore, because the $n = pq$ is known to all as part of the algorithm, the attacker also can determine the costs in launching an attack based on factoring the number n.

Fortunately, there is a great deal of literature on the cost of factoring such a number (Pomerance, 1996). To date, the best-known factoring algorithms are of:

$$O(e^{(((64/9)b)^{(1/3)}(\log b)^{(2/3)})}) \tag{22.1}$$

where b is the number of bits of n expressed as a binary.

Let us consider the range of values for the public key, n, that can be productively used in an attack/defense scenario. First, since p and q, the only prime factors of n, should be approximately the same size, say M decimal digits, then $\log_{10} n \approx 2M$ decimal digits. So, essentially, we can choose n so that $\log_{10} n$ can be any even integer.

Furthermore, in the complexity of the best-known factoring method, the general number field sieve (GNFS), if the runtime of the factoring attack is as in (1) above, then the actual time will be some constant times $e^{(((64/9)b)^{1/3}(\log b)^{2/3})}$, where b is the number of bits of n expressed as a binary.

The constant will be determined by external factors unique to the factoring attempt, such as multithreaded computing, speed of the processor, memory availability, specific implementation of the GNFS, and so on. Nevertheless, relatively speaking, the overall cost of the factoring attempt will not vary greatly compared to the symmetric encryption method mentioned in Section 22.2.

For a specific example, n was factored in a range from 40 to 74 decimal digits (with p and q ranging from 20 to 37 digits) using Wolfram's Mathematica factoring algorithm (described in the Mathematica documentation as "switch[ing] between trial division, Pollard p − 1, Pollard rho, elliptic curve, and quadratic sieve algorithms") (Wolfram, 2014). The factoring times are displayed in Figure 22.1. It is reasonable to select p and q randomly as long as $\log_{10} p \approx \log_{10} q$, as is specified in the RSA algorithm. To demonstrate this principle, 10 different pairs of 30-digit primes p and q were selected randomly, their products n were computed, and each n was factored. The Mathematica runtimes for the 10 samples varied from the mean by more than 40% only once, and none of the other 9 samples varied from the mean by as much as 30%.

If we compute the predicted run times over a large range, using all n values from $10^{70} < n < 10^{350}$, we can predict within a constant the cost of an attack for any value of n in the range. However, by using a best-fit algorithm for $10^{70} < n < 10^{74}$, we can approximate the constant. We thus use this value to predict the cost of an attack over a much larger range.

For large values of $\log_{10} n$, it is more descriptive of the attacker's cost when using a log plot, as seen in Figure 22.2.

For a strong defense, the size of n will certainly need to be greater than 70. However, the function

$$O(e^{(((64/9)b)^{1/3}(\log b)^{2/3})})$$

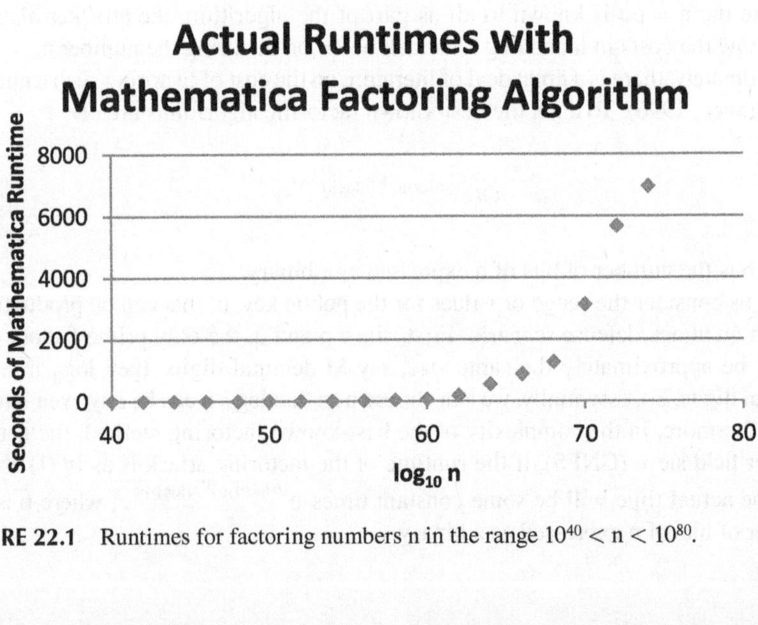

FIGURE 22.1 Runtimes for factoring numbers n in the range $10^{40} < n < 10^{80}$.

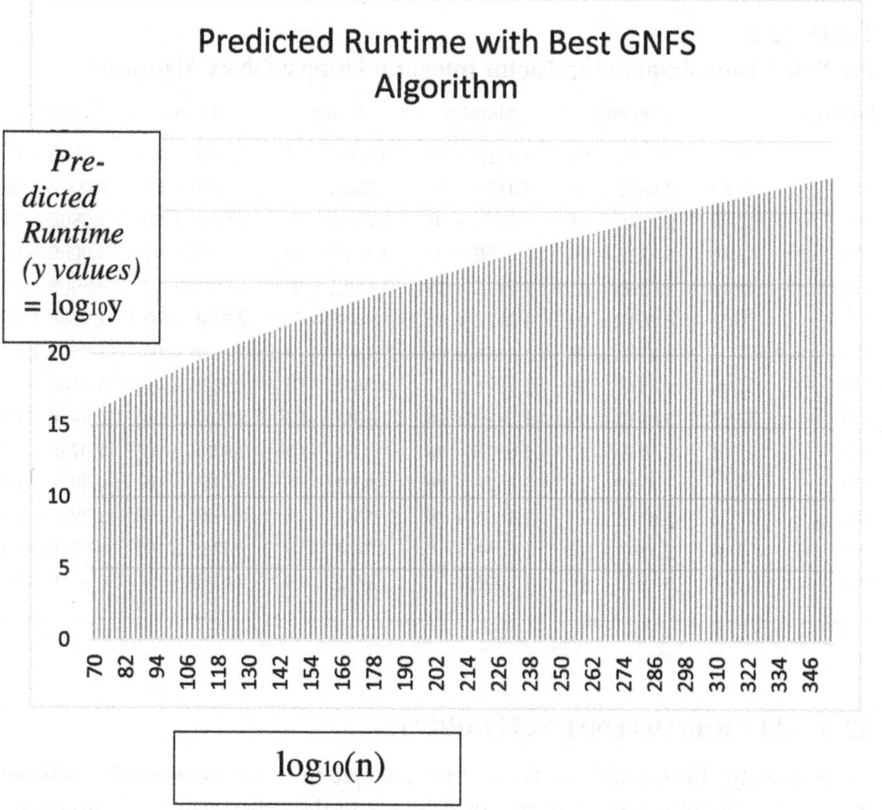

FIGURE 22.2 Predicted runtimes for factoring numbers n in the range $10^{70} < n < 10^{350}$.

grows in faster than polynomial time. So the runtime to factor n will be given by the following values for n as given in Tables 22.1 and 22.2.

This leads to the best-fit linear function: $y = (3.01720 \times 10^{12})x - 2.26540 \times 10^{15}$, with a positive correlation coefficient of 0.95. So this linear fit will give a good (enough) first approximation to the time to factor for any n in a sufficiently large range, as shown in Table 22.1.

TABLE 22.1
Time Required to Factor Integer n Using Mathematica

Decimal Digits of n	x = Mathematica Runtime (Seconds)	y = Predicted Time (Order of Magnitude)
70	3130.66	7.91175E + 15
72	5679.34	1.26921E + 16
74	6967.5	2.0204E + 16

TABLE 22.2
Predicted Time Required to Factor Integer n Using a GNFS Algorithm

log(10) n		Seconds	Minutes	Hours	Days	Years
70	15.9	3.373E + 03	5.622E + 01	9.400E − 01	4.000E − 02	0.000E + 00
80	16.89	2.662E + 04	4.437E + 02	7.390E + 00	3.100E − 01	0.000E + 00
90	17.82	2.172E + 05	3.621E + 03	6.035E + 01	2.510E + 00	1.000E − 02
100	18.68	1.582E + 06	2.636E + 04	4.394E + 02	1.831E + 01	5.000E − 02
110	19.49	1.029E + 07	1.715E + 05	2.858E + 03	1.191E + 02	3.300E − 01
120	20.26	6.064E + 07	1.011E + 06	1.684E + 04	7.018E + 02	1.920E + 00
130	21	3.278E + 08	5.464E + 06	9.107E + 04	3.795E + 03	1.040E + 01
140	21.7	1.643E + 09	2.738E + 07	4.564E + 05	1.902E + 04	5.210E + 01
150	22.37	7.697E + 09	1.283E + 08	2.138E + 06	8.908E + 04	2.441E + 02
160	23.01	3.394E + 10	5.657E + 08	9.428E + 06	3.928E + 05	1.076E + 03
170	23.63	1.417E + 11	2.362E + 09	3.937E + 07	1.640E + 06	4.494E + 03
180	24.23	5.631E + 11	9.386E + 09	1.564E + 08	6.518E + 06	1.786E + 04
190	24.81	2.139E + 12	3.565E + 10	5.941E + 08	2.475E + 07	6.782E + 04
200	25.37	7.793E + 12	1.299E + 11	2.165E + 09	9.019E + 07	2.471E + 05

22.5 ATTACK/DEFENSE SCENARIOS

Unfortunately, this example so far is almost unique in trying to establish a reliable metric in terms of potential attacks or defenses in the cybersecurity environment. The reason for this is that with many other security measures, there is not the same reliability on a specific nature of the attack that might be attempted, and thus the measures cannot be so reliably determined. This is demonstrated in the simple crypto example called P, where one type of attack would surely fail, whereas a different choice of attack would almost always achieve success.

The simplicity of RSA encryption lends itself to the type of analysis described above and the type of metric that can be assigned to it because of a number of characteristics of the RSA. First, RSA is virtually completely scalable, so that a defender establishing an RSA system, as long it is done properly, can make the choice of almost any size integer n as the base of the public information. Second, the choice of n can be taken based on the computational cost of trying to factor that number n, a cost that is very well known and quite stable in terms of research on factoring. Next, despite approximately 40 years of effort in trying to find other ways of breaking RSA, factoring the number n remains the only known way of breaking the cryptosystem. And, finally, it is generally accepted in the number theory community that it is highly unlikely that any vast improvement will be made on factoring research until and unless it is feasible to build and operate robust quantum computers—which many quantum specialists estimate as being several decades in the future. Should quantum encryption become feasible, Shor (1997) has shown how RSA could be broken in such an environment.

Let us consider several examples:

First Scenario: The intent of the Defender is simply to deter the Attacker from any attempt to use a factoring algorithm, because the calculation of the cost will be so high that no rational attacker will bother to attempt the attack, knowing that the cost will be beyond any potential value of the resource to be stolen, or the Attacker's ability to assume the cost for successful attack. Thus, the reasonable attacker will not even bother to try the attack.

The Defender will assess the value of his or her information, and from the analysis choose the appropriate value for n, say n(D). This sets a lower bound for D's actual choice of n. Next, D will estimate A's ability to attack, then convert to a n(A), and choose for the actual n the value so that $n > \max(n(A), n(D))$.

For this scenario, the defender will choose values of n in the range $160 \leq \log_{10} n \leq 200$. The cost involved will deter the attacker even if he or she is using a multipurpose attack with k processors.

Example: The Defender, known as billgates, is aware of press reports that his total worth is $15 billion. Because he is a prudent investor, his holdings are widely distributed to the point that the keys to his accounts are too numerous to remember, so he stores the information in a protected area, guarded by an RSA cryptosystem where the $\log_{10} n = 180$ (from Table 22.2). Since any potential attacker will know that the attack will take over 17,000 years of effort (either in time, money, or cycles), he or she will seek to find another victim.

Second Scenario: The Defender will deliberately choose the cost of the attack to be so low that any and every potential attacker will determine the cost to be acceptable and thus proceed to solve the factoring problem. This would not seem to be the best strategy for the Defender; however, it may be that the Defender has deliberately decided to allow the Attacker to enter because upon retrieving the decrypted information, the Attacker may be led into a trap. This, in cybersecurity technology, is often referred to as a honeypot.

In this case, regardless of the value of D's information, D will estimate A's ability to attack as n(D), then deliberately choose $n < n(D)$.

Third Scenario: The Defender chooses a difficulty level for the RSA that will cause the potential attacker to make a judgment call about whether the cost of an attack would be expected to be less than the projected value of the information obtained in a successful attack. In this way, the Defender and Attacker essentially establish a game theory problem, whereby both the Attacker and Defender will need to establish a cost and benefit of the information being guarded.

Fourth Scenario: A and B use a network for sensitive communication, and they are consistently suspicious of potential attackers attempting to obtain their sensitive information. Both A and B have the ability to regularly change their RSA keys, including the magnitude of these keys. On certain occasions, in order to detect attackers, they lower their level of security to a threshold level, and at this point transmit false information that could lead to the use of such information by an attacker. In this way, the Attacker may be detected.

Fifth Scenario: Reverse Trolling: Trolls are usually thought of as part of an offensive strategy. In this example, the Defender D will "troll" in order to identify or catalog potential opponents. As in the second scenario, D is attempting to direct potential As into a trap area. However, in this case, D with regularity will vary the size of $n(D)$ in order to entice a range of potential attackers and thus hopefully determine different types of lesser or greater attacks with different values of $n(D)$. As before, A and B are using a network, and they can regularly change their RSA keys, and especially lower the security to tempt the third party. Then they might deliberately exchange false information in order to detect one or many Attackers.

22.6 CONCLUSION

Thus, the example described in this chapter hopefully provides a methodology for establishing the strategy for a successful defense strategy based on the precision of the metric that has been designed in this chapter. It is hoped that further investigations in this regard will find other uses for a quantifiable way of estimating the costs of attacks and defenses in a cybersecurity environment.

In essence, then, the use of RSA, under the control of the Defender, can determine precisely the cost to an Attacker of trying to break the encryption. Of course, if the Defender can estimate that the Attacker has the resources to conduct a multiprocessor attack with k processors, then the time to factor indicated by the above can be divided by k.

Why is this advantageous to the Defender? It implies that the Defender can thus issue the challenge, knowing that both Attacker and Defender will be able to calculate the cost of a successful attack.

Given this underlying information, this means that the Defender, in establishing the parameters, can select his or her level of defense to provoke a specific type of response from the Attacker.

The simplicity of RSA encryption lends itself to the type of analysis described above and the type of metric that can be assigned to it because of a number of characteristics of the RSA. First, RSA is virtually completely scalable, so that a defender establishing an RSA system, as long it is done properly, can make the choice of almost any size integer n as the base of the public information. Second, the choice of n can be taken based on the computational cost of trying to factor that number n, a cost that is very well known and quite stable in terms of research on factoring. Next, despite approximately 40 years of effort in trying to find other ways of breaking RSA, factoring the number n remains the only known way of breaking the cryptosystem. And, finally, it is generally accepted in the number theory community that it is highly unlikely that any vast improvement will be made on factoring research until and unless it is feasible to build and operate robust quantum computers—which many quantum specialists estimate as being several decades in the future.

Thus, the example described in this chapter hopefully provides a methodology for establishing the strategy for a successful defense strategy based on the precision of the metric that has been designed in this chapter. It is hoped that further investigations in this regard will find other uses for a quantifiable way of estimating the costs of attacks and defenses in a cybersecurity environment.

REFERENCES

Cohen, F. 1987. Computer viruses: Theory and experiments. *Computers and Security*, 6(1), 22–32.

Diffie, W. and Hellman, M. 1976. New directions in cryptography. *IEEE Transactions on Information Theory*, IT-22, 644–654.

Patterson, W. 2016. *Lecture Notes for CSCI 654: Cybersecurity I.* Howard University, September.

Pomerance, C. 1996. A tale of two sieves. *Notices of the AMS*, 43(12), 1473–1485.

Rivest, R., Shamir, A., and Adleman, L. 1978. A method for obtaining digital signatures and public-key cryptosystems (PDF). *Communications of the ACM*, 21(2), 120–126.

Shor, P. W. 1997. Polynomial-time algorithms for prime factorization and discrete logarithms on a quantum computer. *SIAM Journal on Computing*, 26(5), 1484–1509.

Sone, M. E. and Patterson, W. 2017. *Wireless Data Communication: Security-Throughput Tradeoff.* Saarbrücken, Germany: Editions Universitaires Européennes

Wolfram Corporation. 2014. *Wolfram Mathematica 11.0: Some Notes on Internal Implementation.* Champaign-Urbana, IL, 2014.

PROBLEMS

22.1 Suggest an attack on the 20-letter alphabet encryption method described above that would be an improvement on trying all 20! keys.

22.2 In Chapter 5, we introduced the ABCD method of classifying hackers. Suppose you knew which type of hacker you had to defend against. Using the terminology above, what magnitude of prime products (n) would you choose if you were facing an A, B, C, or D attacker?

22.3 Comment on the five scenarios described above.

23 Behavioral Economics

Behavioral economics is the study of how psychological, cognitive, emotional, cultural, and social factors impact the economic decisions of individuals and institutions, and how these factors influence market prices, returns, and resource allocation. It also studies the impact of different kinds of behavior in environments of varying experimental values. An overview of the topic can be found on Wikipedia (2018).

It is primarily concerned with the rational behavior of economic agents and the bounds of this behavior. In our case, we are interested in the rational behavior of cyberattackers and defenders. Behavioral models typically integrate insights from psychology, neuroscience, and microeconomic theory and, in our case, cybersecurity theory. The study of behavioral economics includes how financial decisions are made.

The use of the term "behavioral economics" in scholarly papers has increased significantly in the past few years, as shown by a recent study. We will consider three prevalent themes in behavioral economics:

- *Heuristics:* Humans make 95% of their decisions using rules of thumb or mental shortcuts.
- *Framing:* The collection of anecdotes and stereotypes that make up the mental emotional filters individuals rely on to understand and respond to events.
- *Market inefficiencies:* These include mispricings and nonrational decision-making.

We have chosen to introduce this topic in a study of the behavioral aspects of cybersecurity because, as we have seen in previous chapters, we can often describe cybersecurity scenarios using an economics model. Among the leading scholars in this new branch of economics are Maurice Allais, Daniel Kahneman, Amos Tversky, Robert Schiller, Richard Thaler, and Cass Sunstein.

In 2002, psychologist Kahneman was awarded the Nobel Memorial Prize in Economic Sciences for "for having integrated insights from psychological research into economic science, especially concerning human judgment and decision-making under uncertainty" (Nobel, 2002). In 2012, economist Schiller received the same Nobel Memorial Prize "for his empirical analysis of asset prices" (Nobel, 2013). In 2017, economist Thaler was also awarded the economics Nobel Memorial Prize for "his contributions to behavioral economics and his pioneering work in establishing that people are predictably irrational in ways that defy economic theory" (Appelbaum, 2017).

23.1 ORIGINS OF BEHAVIORAL ECONOMICS

In the early years of the formal study of economics (the eighteenth century), microeconomics was closely linked to psychology. Microeconomics, as opposed to macroeconomics, deals with the economic decision- making and actions at the level of the individual or small group. Macroeconomics deals with the theories and research on the scale of large organizations or governments. Adam Smith (1759) wrote *The Theory of Moral Sentiments*, which proposed psychological explanations of individual behavior and concerns about fairness and justice. Jeremy Bentham wrote extensively on the psychological underpinnings of utility (Bowring, 1962).

Economic psychology emerged in the twentieth century. Expected utility and discounted utility models began to gain acceptance, generating testable hypotheses about decision-making given uncertainty and intertemporal consumption, respectively. Further steps were taken by Maurice Allais, for example, in setting out the Allais paradox, a decision problem he first presented in 1953 that contradicted the expected utility hypothesis. We will see some of Allais' examples below (Allais, 1953).

In the 1960s, cognitive psychology began to shed more light on the brain as an information processing device (in contrast to behaviorist models). Scholars such as Amos Tversky and Daniel Kahneman began to compare their cognitive models of decision-making under risk and uncertainty to economic models of rational behavior.

Bounded rationality is the idea that when individuals make decisions, their rationality is limited by the tractability of the decision problem, their cognitive limitations, and the time available.

This suggests the idea that humans take shortcuts that may lead to suboptimal decision-making. Behavioral economists engage in mapping the decision shortcuts that agents use in order to help increase the effectiveness of human decision-making. One treatment of this idea comes from Richard Thaler and Cass Sunstein's *Nudge* (2008).

A seminal word in introducing this field appeared in 1979, when Kahneman and Tversky published *Prospect Theory: An Analysis of Decision under Risk*, which used cognitive psychology to explain various divergences of economic decision making from neoclassical theory. Prospect theory has two stages: an editing stage and an evaluation stage (Kahneman and Tversky, 1979). A very good lay description of the work of Kahneman and Tversky appears in the book by Michael Lewis, *The Undoing Project* (2016).

In the editing stage, risky situations are simplified using various heuristics. In the evaluation phase, risky alternatives are evaluated using various psychological principles that include:

- *Reference dependence:* When evaluating outcomes, the decision maker considers a "reference level." Outcomes are then compared to the reference point and classified as "gains" if greater than the reference point and "losses" if less than the reference point. From the perspective of a cyberattacker or cyberdefender, the reference point is estimated as the cost of successfully attacking or defending a cyberenvironment.
- *Loss aversion:* Losses are avoided more than equivalent gains are sought. In a 1992 paper, Kahneman and Tversky found the median coefficient of loss aversion to be about 2.25; that is, losses hurt about 2.25 times more than equivalent gains reward.
- *Nonlinear probability weighting:* Decision makers overweight small probabilities and underweight large probabilities—this gives rise to the inverse-S shaped "probability weighting function."
- *Diminishing sensitivity to gains and losses:* As the size of the gains and losses relative to the reference point increase in absolute value, the marginal effect on the decision maker's utility or satisfaction falls.

23.2 UTILITY

Utility is a most interesting word in the English language, with many different connotations. In our case, it will not refer to the backup shortstop on a baseball team, nor the company that provides the distribution of electrical services to a community.

In economics generally and behavioral economics in particular, the concept of utility is used to denote worth or value. The term was introduced initially by two of the founders of economics in the mid-nineteenth century, Jeremy Bentham and John Stuart Mill (1863), and it meant, in their context, that the theory of utilitarianism describes a measure of satisfaction relating to economic value. In modern economic theory, the term *utility* has evolved to mean a function that represents a consumer's preference for a decision made over a set of possible economic choices.

If you have a set of potential economic alternatives and you have a way of ordering these preferences in your favor, a utility function can represent those preferences if it is possible to assign a number to each alternative; then, also, you can say that alternative a is assigned a number greater than alternative b, if and only if you prefer alternative a to alternative b.

When you have this situation over perhaps a finite number of choices, the individual selecting the most preferred alternative is necessarily also selecting the alternative that maximizes this associated utility function.

Consider creating an example that will illustrate the thinking about maximizing utility over the first century-plus of utility theory. This example will hopefully lead us to understanding the contradiction that has arisen in this classical approach to understanding economic choices.

Suppose that we value a number of commodities at different levels. For the sake of the example, let us choose pairs of shoes, and let's say that we value them as follows (Figure 23.1):

Flip-Flops	$10
Sandals	$20
Sneakers	$30
Overshoes	$40
Dress Shoes	$50

Classical utility theory concludes that if you have a range of choices of different values, and every one of the choices increases in value by the same amount, the set of choices would remain the same. Going to our example in the chart, if a consumer decided that the greatest value to him or her would be the $30 sneakers, then if the shoe store next door to the original one was selling sneakers at the same prices also offering a free $30 gift bag, then the customer would still choose the sneakers among the other choices.

We have simplified choices by assigning an actual monetary value to the choice of each product. In general, we would describe the utility by a function u such as u(Flip-flops), u(Sandals), and so on. Thus for this example, u(Flip-flops) = $10, u(Sandals) = $20, and so on.

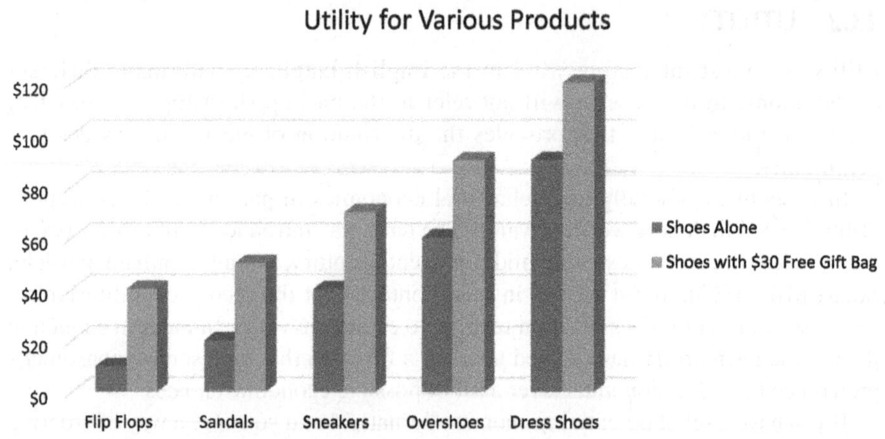

FIGURE 23.1 Illustrating a fundamental principle of utility theory.

23.3 ALLAIS'S CHALLENGE TO UTILITY THEORY

However, the French researcher Maurice Allais (1953) conducted a number of experiments that belied and have basically overthrown the classical theory of utility. Furthermore, Allais's research has been supported and confirmed by many other scholars, perhaps most notably the Israeli scientists Daniel Kahneman and Amos Tversky.

Here are two examples from the original paper by Allais as described by Kahneman and Tversky (1979).

We first show that people give greater weight to outcomes that are considered certain relative to outcomes that are merely probable (the certainty effect). Here are two choice problems modeled after Allais's example. The number of respondents in the Kahneman-Tversky repeat of Allais's research who answered each problem is denoted by N, and the percentage who chose each option is given in brackets.

Problem 23.1 Choose between Choice A and B where you would receive a payoff in a given amount, subject to the corresponding probability.

Choice A		Choice B	
Payoff	**Probability**	**Payoff**	**Probability**
$2500	0.33	$2400	1.00
$2400	0.66		
$0	0.01		
Result of testing the experiment with N = 72 subjects:			
18%		72%	

Problem 23.2 Choose between:

Choice C		Choice D	
Payoff	Probability	Payoff	Probability
$2500	0.33	$2400	0.34
$0	0.67	$0	0.66
Result of testing the experiment with N = 72 subjects:			
83%		17%	

We will show the paradox demonstrated by these results. We assume that the specific choice chosen more often is indicated by the majority of results for the N subjects. In Problem 23.1, 18% of the subjects chose A, and 82% chose B. But for the same subjects, when asked to choose for Problem 23.2, 83% of the subjects chose C, and 17% chose D.

Clearly the subjects picked Choice B from Problem 23.1 by a substantial majority. The same respondents, when given Problem 23.2, were even more definitive in choosing Choice C over Choice D.

This pattern of preferences violates expected utility theory in the manner originally described by Allais.

According to that theory, with $u(0) = 0$, Problem 23.1 implies, since Choice B is preferred by the subjects over Choice A:

$$0.33u(2500) + 0.66u(2400) < u(2400)$$
$$\Rightarrow 0.33u(2500) > 0.34u(2400)$$

while the preference in Problem 23.2 implies the reverse inequality:

$$0.33u(2500) < 0.34u(2400)$$

Note that Problem 23.2 is obtained from Problem 23.1 by eliminating a .66 chance of winning $2400 from both prospects under consideration. In other words, the representation of the choices from A and B to C and D is a constant. Classical utility theory would say that since the relationship between the choices has not changed, the preference for the outcome would not change. But this change produces a greater reduction in desirability when it alters the character of the prospect from a sure gain to a probable one than when both the original and the reduced prospects are uncertain. A simpler demonstration of the same phenomenon involving only two-outcome gambles is given below. This example is also based on Allais.

Problem 23.3 Choose between:

Choice A		Choice B	
Payoff	Probability	Payoff	Probability
$4000	0.80	$3000	1.00
$0	0.20		
Result of testing the experiment with N = 95 subjects:			
20%		80%	

Problem 23.4 Choose between:

Choice A		Choice B	
Payoff	**Probability**	**Payoff**	**Probability**
\$4000	0.20	\$3000	0.25
\$0	0.80	\$0	0.75
Result of testing the experiment with N = 95 subjects:			
65%		35%	

Again, according to classic utility theory, with u(0) = 0, Problem 23.3 implies, since Choice B is preferred by the subjects over Choice A:

$$0.80u(4000) < u(3000)$$

while the preference in Problem 23.4 implies the reverse inequality:

$$0.20u(4000) > 0.25u(3000) \Rightarrow 0.80u(4000) > u(3000)$$
(multiply each side by 4)

In this pair of problems, as well as in all other problem pairs in this section, almost half the respondents (45%) violated expected utility theory by flipping from one side to the other. To show that the modal pattern of preferences in Problems 23.3 and 23.4 is not compatible with the theory, set u(0) = 0, and recall that the choice of B implies u(3000)/u(4000) >4/5, whereas the choice of C implies the reverse inequality.

The examples of Problems 23.5 and 23.6 demonstrate that the same effect, the certainty effect, even when the outcomes are nonmonetary.

Problem 23.5 Choose between Choice A and B where you would receive a nonmonetary payoff in a given amount (even though the subjects could perform a conversion to a monetary value), subject to the corresponding probability.

Choice A		Choice B	
Payoff	**Probability**	**Payoff**	**Probability**
Three-week tour of England, France, Italy	0.50	One-week tour of England	1.00
No tour	0.50	No tour	0.00
Result of testing the experiment with N = 72 subjects:			
22%		78%	

Problem 23.6 Choose between Choice A and B where you would receive a nonmonetary payoff, subject to the corresponding probability.

Choice A		Choice B	
Payoff	Probability	Payoff	Probability
Three-week tour of England, France, Italy	0.05	One-week tour of England	0.10
No tour	0.95	No tour	0.90
Result of testing the experiment with N = 72 subjects:			
67%		33%	

The same effect is also demonstrated when there is a significant probability of an outcome as opposed to when there is a possibility of winning, even though extremely remote. Where winning is possible but not probable, most people choose the prospect that offers the larger gain.

Problem 23.7 Choose between Choice A and B where you would receive a payoff in a given amount, subject to the corresponding probability.

Choice A		Choice B	
Payoff	Probability	Payoff	Probability
\$6000	0.45	\$3000	0.90
\$0	0.55	\$0	0.10
\$0	0.01		
Result of testing the experiment with N = 66 subjects:			
14%		86%	

Problem 23.8 Choose between Choice A and B where you would receive a payoff in a given amount, subject to the corresponding probability.

Choice A		Choice B	
Payoff	Probability	Payoff	Probability
\$6000	0.001	\$3000	0.002
\$0	0.999	\$0	0.998
Result of testing the experiment with N = 66 subjects:			
73%		27%	

Note that in Problem 23.7, the probabilities of winning are substantial (0.90 and 0.45), and most people chose the prospect where winning is more probable. In Problem 23.8, there is a possibility of winning, although the probabilities of winning are minuscule (0.002 and 0.001) in both prospects. In this situation where winning is possible but not probable, most people chose the prospect that offers the larger gain.

23.4 APPLICATION OF THE ALLAIS-KAHNEMAN-TVERSKY APPROACH TO CYBERSECURITY

Consider now the development of similar choice problems set in the cybersecurity environment. Suppose we have an attacker who has identified as a target for a ransomware attack a potential victim for which a hacker can assign a monetary value as a ransom.

The attacker can launch a ransomware attack for which he or she could estimate the likelihood of obtaining the ransom. He or she could configure the attack to request a ransom of either $2400 or $2500, and estimates that the lower level of ransom will, on the one hand (Choice A), succeed 66% of the time for the $2400 ransom, but only 33% of the time for the larger ransom, and fail completely 1% of the time. Or, the attacker's second choice would be to simply employ the $2400 ransom request, which it is felt would succeed 100% of the time.

Problem 23.9 Choose between Choice A and B where you would receive a payoff in a given amount, subject to the corresponding probability.

Choice A		Choice B	
Payoff	**Probability**	**Payoff**	**Probability**
$2500	0.33	$2400	1.00
$2400	0.66		
$0	0.01		

On the other hand, the attacker could make the following choices:

Problem 23.10 Choose between:

Choice C		Choice D	
Payoff	**Probability**	**Payoff**	**Probability**
$2500	0.33	$2400	0.34
$0	0.67	$0	0.66

The numbers may seem familiar from the example in Section 23.3, and of course this is not by accident. Although this experiment has not yet been conducted (to our knowledge), if the results would be similar to the Allais-Kahneman-Tversky data, this would be very useful in comparing cybersecurity economic behavior to the more general research on microeconomics in the context of what we are learning in behavioral economics.

23.5 NUDGE

Nudging is a subtle (or perhaps "stealth") mechanism to influence behavior as opposed to direct education or legislation. Nudge is a concept in behavioral science, political theory, and economics that proposes positive reinforcement and indirect suggestions as ways to influence the behavior and decision-making of groups or individuals. Nudging contrasts with other ways to achieve compliance. The concept has influenced British and American politicians. Several nudge units exist around the world at the national level (UK, Germany, Japan, and others) as well as at the international level (OECD, World Bank, UN).

The first formulation of the term and associated principles was developed in cybernetics by James Wilk before 1995 and described by Brunel University academic D. J. Stewart as "the art of the nudge," sometimes referred to as micronudges (Potter, 2018). It also drew on methodological influences from clinical psychotherapy tracing back to Gregory Bateson. The nudge has been called a "microtargeted design" geared toward a specific group of people.

Nudge theory was perhaps first brought to prominence with Richard Thaler and Cass Sunstein's book *Nudge: Improving Decisions About Health, Wealth, and Happiness* (2008). It also gained a following among American and British politicians in the private sector and in public health. The authors refer to influencing behavior without coercion as libertarian paternalism and the influencers as choice architects. Thaler and Sunstein defined their concept as:

> A nudge, as we will use the term, is any aspect of the choice architecture that alters people's behavior in a predictable way without forbidding any options or significantly changing their economic incentives. To count as a mere nudge, the intervention must be easy and cheap to avoid. Nudges are not mandates. Putting fruit at eye level counts as a nudge. Banning junk food does not.

Drawing on behavioral economics, the nudge is more generally applied to influence behavior. Here are a few examples of the results of the application of "nudge theory":

- One of the most frequently cited examples of a nudge is the etching of the image of a housefly into the men's room urinals at Amsterdam's Schiphol Airport, which is intended to "improve the aim." Men now had something to aim for—even subconsciously—and spillage is reduced by 80%.
- Sunstein and Thaler have urged that healthier food be placed at sight level in order to increase the likelihood that a person will opt for that choice instead of less healthy option.
- Similarly, another example of a nudge in influencing a positive outcome is switching the placement of junk food in a store, so that fruit and other healthy options are located next to the cash register, while junk food is relocated to another part of the store.
- Countries where people have to opt in to donating organs generally see a maximum of 30% of the population registering to donate. In countries where people are automatically enrolled in organ donation schemes and have to actually opt out, only about 10% to 15% of people bother opting out, providing a far larger pool of organ donors.
- The American grocery store Pay & Save placed green arrows on the floor leading to the fruit and vegetable files. They found shoppers followed the arrows 9 times out of 10—and their sales of fresh produce skyrocketed.

- In some schools, the cafeteria lines are carefully laid out to display healthier foods to the students. In an experiment to determine its effect, it was shown that students in the healthy lines make better food choices, with sales of healthy food increasing by 18%.
- In the UK, people in arrears on their taxes were sent reminders that were worded using social normative messages. Phrases such as "9 out of 10 people in your area are up to date with tax payments." By making them seem like the outliers, tax payments from people who were sent those letters was 15% up compared to the norm.

In 2008, the United States appointed Sunstein, who helped develop the theory, as administrator of the Office of Information and Regulatory Affairs. Notable applications of nudge theory include the formation of the British Behavioural Insights Team in 2010, often called the "Nudge Unit," at the British Cabinet Office, headed by David Halpern. Both Prime Minister David Cameron and President Barack Obama sought to employ nudge theory to advance domestic policy goals during their terms. In Australia, the government of New South Wales established a Behavioural Insights community of practice. There are now more than 80 countries in which behavioral insights are used.

23.6 AN APPLICATION OF NUDGE THEORY TO CYBERSECURITY

Suppose there are 10 files, each containing some information of value, perhaps, for example, the location of a bank account. All of these files are encrypted on the host's system. A cyberattacker is able to penetrate the system and determines that he or she has enough time before being detected to obtain the information from at least one of the 10 encrypted files. For 9 of these 10 values, the file sizes are between 1 and 5 kB.

The 10th file, as established by the defender, is information of an economic value of only, say, $5.00. But on the other hand, the file is padded before printing so that the file size becomes 25 kB.

When the attacker finds the 10 encrypted files, he or she may be nudged toward the 10th file, of 25 kB, because the clear perception of the file size being greater than all the others would suggest the potential of greater value in successfully attacking that file. Of course, being nudged toward the 25 kB file will cause the attacker to have to spend considerably more effort to decrypt the file, thus increasing exposure to detection and capture.

23.7 MAXIMIZERS OR SATISFICERS?

Psychologists tell us approaches to decision-making tend to fit into one of two categories: you are either a *maximizer* or a *satisficer*. A maximizer is a person who strives to make a choice that will give them the maximum benefit later on. A satisficer is a person whose choices are determined by more modest criteria.

Maximizers task themselves with making the most informed, intelligent decisions. Thus, we might expect that the outcome of their approach would be superior decisions. But this assumption has been contradicted by numerous studies, which have found that maximizers are often less effective in a decision-making environment and suffer under the pressure of high self-expectations. Setting unachievable goals may impede the ultimate goal when making choices.

23.7.1 Satisficers

The "satisficing" concept was first proposed by Nobel economist Herbert A. Simon, who created the blended word by combining "satisfying" and "sufficing." He developed the idea as a way of explaining a particular form of decision-making or cognitive heuristic (Simon, 1958). He believed that when satisficers are presented with a decision to make, they will consider what they want to gain or preserve from a situation, then evaluate their options to find the solution that meets their requirements.

When choosing which car to purchase, for instance, Mary the satisficer will consider the use of the vehicle (for a long commute to work) and decide that she would like this new car to be fuel efficient. She may also decide that she would like the seats to be heated, and then she looks at three cars for sale:

Car 1: A new car with heated seats, but low fuel efficiency.
Car 2: An older car that is fuel efficient and has heated seats.
Car 3: A new car that is fuel efficient and boasts heated seats and a spacious interior, at little more than the cost of car 2.

Mary will discount Car 1, as the fuel costs will be too high for the long commute to work. With a choice between Cars 2 and 3, she might well reject the benefit of the additional space in the third car as being unnecessary and a lavish additional expense, instead opting to buy Car 2, as it meets her initial decision-making criteria.

23.7.2 Maximizers

Like satisficers, maximizers refine their options to those that will fulfill their essential needs when making a decision. But they will subsequently pursue the option that will provide them with the maximum benefit or highest utility. Had Mary, from our previous example, been a maximizer, she would likely have wanted to pay a little extra money to buy Car 3 for its extra space, regardless of whether she really needed the additional room for her commute.

Maximizers will set themselves high standards during decision-making and will aim for them but are often disappointed when they fail to achieve them, dwelling on what they have missed out on rather than what they have. By contrast, a satisficer will be satisfied with the option chosen even if it was not the best option wished for.

As a comment related to our earlier sections on gender, there is research suggesting that "the concept of satisficing … is a more appropriate way to view women's working lives than are either choice or constraint theories" (Corby and Stanworth, 2009).

23.8 BOUNDED RATIONALITY

Bounded rationality is the idea that when individuals make decisions, their rationality is limited by the difficulty of the decision problem, the cognitive limitations of their minds, and the time available to make the decision. Decision-makers in this view act as satisficers, as defined above, seeking a satisfactory solution rather than an optimal one. Simon proposed bounded rationality as an alternative basis for the mathematical modeling of decision-making, as used in economics, political science, and related disciplines. It complements "rationality as optimization," which views decision-making as a fully rational process of finding an optimal choice given the information available. Simon used the analogy of a pair of scissors, where one blade represents "cognitive limitations" of actual humans and the other the "structures of the environment," illustrating how minds compensate for limited resources by exploiting known structural regularity in the environment.

This implies the idea that humans take reasoning shortcuts that may lead to suboptimal decision-making. Behavioral economists engage in mapping the decision shortcuts that agents use in order to help increase the effectiveness of human decision-making. One treatment of this idea comes from nudge theory, as discussed above. Sunstein and Thaler recommend that choice architectures be modified in light of human agents' bounded rationality.

Applications for behavioral economics include the modeling of the consumer decision-making process for applications in artificial intelligence and machine learning.

REFERENCES

Allais, M. 1953. Le comportement de l'Homme Rationnel devant le Risque, Critique des Postulats et Axiomes de l'Ecole Americaine. *Econometrica*, 21, 503–546.

Appelbaum, B. 2017. Nobel in economics is awarded to Richard Thaler. *The New York Times*, 2017-10-09. https://www.nytimes.com/2017/10/09/business/nobel-economics-richard-thaler.html

Bowring, J. (ed.). 1962. The works of Jeremy Bentham, London, 1838–1843. In *Internet Encyclopedia of Philosophy*. Reprinted New York. https://www.iep.utm.edu/bentham/#SH6a

Corby, S. and Stanworth, C. 2009. A price worth paying?: Women and work—Choice, constraint or satisficing. *Equal Opportunities International*, 28(2), 162–178. https://doi.org/10.1108/02610150910937907

Kahneman, D. and Tversky, A. 1979. Prospect theory: An analysis of decision under risk. *Econometrica. The Econometric Society*, 47(2), 263–291. doi: 10.2307/1914185. JSTOR 1914185.

Lewis, M. 2016. *The Undoing Project*. New York: W. W. Norton.

Mill, J. S. 1863. Utilitarianism. https://www.utilitarianism.com/mill1.htm

Nobel Prize. 2002. The Sveriges Riksbank Prize in Economic Sciences in Memory of Alfred Nobel 2002: Daniel Kahneman. https://www.nobelprize.org/prizes/economic-sciences/2002/summary/

Nobel Prize. 2013. The Sveriges Riksbank Prize in Economic Sciences in Memory of Alfred Nobel 2013: Robert J. Shiller. https://www.nobelprize.org/prizes/economic-sciences/2013/summary/

Potter, R. 2018. Nudging—The practical applications and ethics of the controversial new discipline. *The Ecoonomics Review at NYU*, March 23. https://theeconreview.com/2018/03/23/nudging-the-practical-applications-and-ethics-of-the-controversial-new-discipline/

Simon, H. A. 1958. Rational choice and the structure of the environment. *Psychological Review*, 63(2), 129–138.

Smith, A. 1759. The Theory of Moral Sentiments, http://www.earlymoderntexts.com/assets/pdfs/smith1759.pdf
Thaler, R. and Sunstein, C. 2008. *Nudge: Improving Decisions about Health, Wealth, and Happiness*. New Haven, CT: Yale University Press.
Wikipedia. 2018. Behavioral Economics. https://en.wikipedia.org/wiki/Behavioral_economics

PROBLEMS

23.1 Find your own sample of students to conduct the four Allais-Kahneman-Tversky choice problems. Compare your results to the published data.

23.2 Should one expect the aforementioned results to change depending on the sector of the economy utilized, for example, the cyberenvironment, for one case?

23.3 Find a number of examples of the application of nudge theory.

24 Fake News

Clearly, to date, the most publicized set of examples of "fake news" deal with the multiple events that have occurred over a number of years, but most recently the events leading to and since the United States 2016 presidential election.

There should be an increasing focus on strategies to identify the intrusions that we might label fake news and the development of techniques to identify and thus defeat such practices. Certainly major technology corporations such as Facebook and Yahoo have been investing in research to try to address these problems.

However, because there is no clear methodology or set of methodologies that can be employed to completely or even partially eliminate this problem, we will try in this section to identify some partial approaches to determining the validity of news and to suggest certain measures in order to eliminate such the fake news practice.

24.1 A FAKE NEWS HISTORY

It should first be recognized that fake news has indeed a long history, just as in our earlier chapter on the history of cryptology, we noted the involvement of Julius Caesar in an early cryptographic method. Certainly historically related to the Emperor Caesar was the person at least indirectly involved in his assassination: Mark Antony. Antony himself became the victim of fake news, which resulted in his eventually committing suicide (A&E, 2017; MacDonald, 2017).

Moving ahead many centuries, leading up to the American Revolution, Benjamin Franklin published fake news about Native Americans allied with King George III involved in "scalping" in order to gain support for the forthcoming revolution (Harrington, 2014).

During the period of slavery in the United States, its supporters developed many fake news stories about African-Americans, both about purported slave rebellions and also stories of African-Americans spontaneously turning white, all of which brought fear to many whites, especially in the South (Theobald, 2006).

At the end of the nineteenth century, led by Joseph Pulitzer and other publishers and usually referred to as "yellow journalism," writers pushed stories that led the United States into the Spanish-American War when the USS Maine exploded in the Havana, Cuba, harbor (Soll, 2016).

In 1938, the radio drama program "The Mercury Theater on the Air," directed by Orson Welles, aired an episode called "War of the Worlds," simulating news reports of an invasion of aliens in New Jersey. Before the program had begun, listeners were informed that this was just a dramatization. However, most listeners missed that part and therefore believed that the invasion was real. One concrete result was that the attack, supposedly in Grover Mill, New Jersey, resulted in residents attacking a water tower because the broadcaster identified it falsely as alien (Chilton, 2016).

Another example that might very well be considered fake news is cited in the earlier chapter on steganography. In 1947, FBI Director J Edgar Hoover described, in a *Reader's Digest* article, microdot technology as "the enemy's masterpiece of espionage" (Hoover, 1946). However, as was later noted in a commentary on Hoover's article, he had attributed the invention of microdots to "the famous Professor Zapp at the Technical University Dresden." However, there never was a Professor Zapp at that university, and microdot historian William White (1992) denounced Hoover's article as a "concoction of semitruths and overt disinformation."

Of course, all the examples cited above occurred before the existence of cybertechnology. The impact of the internet on the promulgation of fake news is such that the implementation and spread of such information can move instantly and affect internet users by the millions.

24.2 FAKE NEWS RESURGENCE, ACCELERATION, AND ELECTIONS

Over the last decade, the use of fake news has been applied in a number of areas: for financial gain, for political purposes, for amusement, and for many other reasons.

There has now been a great deal written about the use of such fake news in order to influence not only the 2016 United States presidential election, but also elections in numerous other countries throughout the world. One especially egregious example was the result of the news story usually called the "Pizzagate" conspiracy theory, which accused a certain pizzeria in Washington, DC, as hosting a pedophile ring run by the Democratic Party. In December 2016, an armed North Carolina man, Edgar Welch, traveled to Washington, DC, and opened fire at the Comet Ping Pong pizzeria identified in the Pizzagate fake news. Welch pleaded guilty to charges of unlawful interstate transport of firearms and assault with a deadly weapon and was sentenced to 4 years in prison (Kang, 2016).

During the time of the 2016 U.S. election, it was discovered that many of the fake news stories being promulgated originated in a single city, Veles in Macedonia, with numerous fake news organizations employing hundreds of teenagers to produce and distribute sensationalized fake news stories for different U.S.-based companies and parties (Kirby, 2016).

24.3 WHAT IS FAKE NEWS?

As we have seen, there is nothing new about fake news. Most of us first encounter this phenomenon the first time we discover a lie. However, it can fairly be argued that the telling of a lie presented as part of the technology may seem so impressive as to convince us of the truthfulness of the message because of the elaborate wrapping.

It is really only in the internet age that the toolset has become readily available to create a very (seemingly) realistic message that will fool many readers.

The very rapid expansion in the past few years of fake news sites can be attributed to many factors. Clearly, the use of sophisticated composition software to produce websites with very sophisticated appearances provides the opportunity to create such fake news sites.

However, the motivation for creating such sites is undoubtedly a reason for the rapid expansion. There is a financial incentive for many of these sites, either because their creators are being paid to produce them (note the example of Veles, Macedonia, cited above) or because the site itself may provide a mechanism for readers to buy—whether a legitimate product or a scam.

It seems that more sites may have as a more central purpose the advocacy of a political point of view, which may put forward completely false information or information with a mix of truth and half-truth.

24.4 SATIRE OR FAKE NEWS?

A difficult challenge for the consumers of various information is that there may be substantial similarity between sites that might be considered fake news and those that are meant to be satirical. Certainly, satire is an important form of storytelling and criticism.

Indeed, it is well known that many (if not all) nursery rhymes that we know from childhood are actually satire disguised as children's tales to avoid retribution since the satirical meaning may actually constitute criticism of a powerful monarch. Consider these examples (Fallon, 2014):

Baa Baa Black Sheep

Baa, baa, black sheep, have you any wool?
Yes sir, yes sir, three bags full!

One for the master,
One for the dame,
And one for the little boy
Who lives down the lane.

It is thought that this poem dates back to feudal times and the institution of a harsh tax on wool in England. One-third would be taken for the king and nobility, and one-third for the church, leaving little for the farmers.

Little Jack Horner

Little Jack Horner
Sat in the corner,
Eating a Christmas pie;
He put in his thumb,
And pulled out a plum,
And said, "What a good boy am I!"

One interpretation has "Little Jack" standing in for Thomas Horner. Horner was a steward who was deputized to deliver a large pie to Henry VIII concealing deeds to a number of manors, a bribe from a Catholic abbot hoping to save his monastery from the king's anti-Catholic crusade. Horner ended up with one of the manors for himself, and it's believed by many that he reached into the pie and helped himself to a deed, though his descendants have insisted that he paid for the property fair and square.

Yankee Doodle

Yankee Doodle went to town
Riding on a pony
Stuck a feather in his cap
And called it macaroni!

It's clear that Yankee Doodle is a silly figure in this classic ditty, which dates back to the Revolutionary era. At the time, however, macaroni was the favored food of London dandies, and the word had come to refer to the height of fashion. British soldiers, who originally sang the verse, were insulting American colonists by implying they were such hicks they thought putting feathers in their hats made them as hip and stylish as London socialites.

There is a major effort now from many quarters in trying to identify techniques to be able to classify websites for social media in terms of their "fakeness." The major companies hosting social media, for example, Facebook and Yahoo, have initiatives to identify and disqualify fake news accounts.

It is generally felt that automated approaches using artificial intelligence techniques are probably needed to support such efforts, but this is a complex technical issue and indeed may be unsolvable.

24.5 DISTINGUISHING SATIRE FROM FAKE NEWS

In the next section, we will identify a number of tests that can be applied to assist in determining the status of a questionable website or Facebook message we might encounter. But here, we will analyze a few chosen examples to try to determine their validity. As we have indicated, there has been an explosion in the number of fake news sites. At the time of this writing, a Wikipedia article identifies 77 from U.S. sources—and in the Philippines alone, another 84 sites (Wikipedia, 2018).

What follows are set of examples of either "fake news" or some we might designate "not fake news."

24.5.1 DailyBuzzLive.com

DailyBuzzLive is an online magazine or e-zine that specializes in sensational articles. The flavor can be ascertained from some of the sections of the zine indicated by the menu selections on the homepage: "Controversial," "Viral Videos," "Weird," "Criminal," "Bad Breeds."

One immediate clue as to its validity is that there is no obvious way of determining the publisher. The topics vary widely, but a few sample headlines include:

USDA Allows US to be Overrun With Contaminated Chicken from China
Human Meat Found In McDonald's Meat Factory
People Call for Father Christmas to be Renamed 'Person Christmas'

With respect to the last article regarding Christmas, there is no author indicated, nor any date on the article. From the article:

> For a lot of people, the first question which springs to mind is: Are we really living in a world where we're unable to refer to a Saint and more importantly, a much-loved traditional festive figure as male for fear of offending a portion of society?
>
> It turns out some people think it's a good idea, after all this year has seen companies like John Lewis get rid of boys and girls labels in their children's clothes.

This article "People Call for Father Christmas to be Renamed 'Person Christmas'" might be viewed by some as merely satire. However, the tone is highly political as a slam at the LGBT community. Although there is some legitimate quoting in the article going back to the third century AD, there are a number of indicators that this would qualify as fake news. First, although this is a very lengthy article, there is no author cited anywhere in the article. Furthermore, the headline, beginning with "People Call for…" never refers to anyone actually making that "call."

24.5.2 ABCNEWS.COM.CO

This website ABCnews.com.co is no longer in existence. If you enter that URL, you will find the statement that is common to nonexistent sites that begins with "Related Links." It was apparently owned by the late Paul Horner.

This one is easy to detect. In fact, the original website has already been taken down. The clue is that the URL, although appearing to be the website for ABC News (ABCnews.com), actually ends with the Internet country code ".co" for Colombia in South America. This trick has been used by a number of other fake news sites.

24.5.3 THEONION.COM

On first glance, one might consider *The Onion* fake news. However, although not doing so on the masthead on top of the website, *The Onion* identifies itself as satire, and it follows a lengthy tradition in this genre, much like the Mother Goose rhymes above with a political impact. *The Onion* is unlike the *Daily Buzz* example above, which does not identify itself as satire.

Several of the Onion articles, clearly satirical in nature, are:

Nation Not Sure How Many Ex-Trump Staffers It Can Safely Absorb

New Ted Cruz Attack Ad Declares Beto O'Rourke Too Good for Texas

Elizabeth Warren Disappointed After DNA Test Shows Zero Trace of Presidential Material

The Onion on its website has a section (https://www.theonion.com/about) that indicates, in a rather obvious satirical fashion, the supposed history of its publication:

> *The Onion* is the world's leading news publication, offering highly acclaimed, universally revered coverage of breaking national, international, and local news events. Rising from its humble beginnings as a print newspaper in 1756, *The Onion* now enjoys a daily readership of 4.3 trillion and has grown into the single most powerful and influential organization in human history.
>
> In addition to maintaining a towering standard of excellence to which the rest of the industry aspires, *The Onion* supports more than 350,000 full- and part-time journalism jobs in its numerous news bureaus and manual labor camps stationed around the world, and members of its editorial board have served with distinction in an advisory capacity for such nations as China, Syria, Somalia, and the former Soviet Union. On top of its journalistic pursuits, *The Onion* also owns and operates the majority of the world's transoceanic shipping lanes, stands on the nation's leading edge on matters of deforestation and strip mining, and proudly conducts tests on millions of animals daily.

In addition, in the same section, there is a listing of the editorial staff. The editor-in-chief, Chad Nackers, can also be found through Wikipedia, and it is clear that he is a real person and the actual editor.

24.5.4 Infowars.com

This site is closely affiliated with Alex Jones, who has long been identified as a conspiracy theorist, ranging from such conspiracies as the "birther conspiracy" about former President Obama to the argument that school shootings such as Sandy Hook and Lakeland were faked, to the "Pizzagate" story cited earlier in this chapter. However, beyond that connection, the other giveaway is the headline photograph of what is entitled "Invasion Begins! Migrant Caravan Arrives At US/Mexico Border" showing immigrants scaling a wall, presumably to enter the United States (https://www.infowars.com/invasion-begins-migrant-caravan-arrives-at-us-mexico-border/). When one checks other media sources, at the time this article was published (November 15, 2018), a vast number of reports from other sources indicated that the "caravan" of refugees (so designated by President Trump during the 2018 election campaign) was hundreds of miles from the United States border.

24.5.5 New Yorker

Andy Borowitz has written a series of satirical articles for several years in the *New Yorker* (https://www.newyorker.com/contributors/andy-borowitz). He makes it very clear in the headline over his articles that they are "Satire from the Borowitz Report"; in addition, also above his articles is a line indicating they are "Not the News." Borowitz created the Borowitz Report in 2001.

Some sample titles are:

> John Kelly Departs White House with Nuclear Codes Hidden in Pants
>
> Trump Warns That Florida Recount Could Set Dangerous Precedent of Person with Most Votes Winning
>
> Trump Now Says If Migrants Throw Rocks Military Will Respond with Paper or Scissors

24.5.6 EmpireNews.net

In one alarming story, the person indicated in the headline, Jerry Richards, is alleged to have murdered over 700 people in Naples, Florida. In this case, since allegedly the murderer has been arrested, there should be court records as well as other news articles about this event. There are not.

At least, in an "About" section, *Empire News* indicates:

> Empire News is intended for entertainment purposes only. Our website and social media content uses only fictional names, except in cases of public figure and celebrity parody or satirization. Any other use of real names is accidental and coincidental.
>
> The content of this website, including graphics, text, and other unspecified elements are all © Copyright 2014–2015 Empire News LLC. Content may not be reprinted or re-transmitted, in whole or in part, without the express written consent of the publisher.

24.5.7 BeforeItsNews.com

The November 12, 2018, article "Operation Torch California" (https://beforeitsnews.com/v3/terrorism/2018/2461448.html) regarding the devastating forest fire both in Northern and Southern California raised a number of questions. The first warning sign occurred with the quote in the opening: "*Operation Torch California* is a very real ongoing black operation being conducted by the U.S. Intelligence Community … These false flag terrorist attacks are first and foremost a highly sophisticated psyop." The quote was attributed to an unnamed "Intelligence Analyst & Former U.S. Military Officer." Next, there are references to the acronym DEWs, which is never defined. (Could that mean digital electronic warfare systems, drought early warning systems, or something else?)

From that point on, the article bounces from one wild statement to another, linking this story with Hurricane Michael in Florida, to ISIS and Al-Qaeda, to aluminum oxide from coal fly ash. Among many other wild statements, it is alleged that "To name the most devastating fire in California history *Camp Fire* represents the profound cynicism associated with this well-planned pyro-psyop. How easy it is to now blame that geoengineered wildfire on a simple 'camp fire'." It was elsewhere reported that the Camp Fire, the deadliest and most destructive wildfire in California history, was called the Camp Fire because it originated at Camp Creek Road.

24.5.8 Centers for Disease Control

This is perhaps the most perplexing of all the examples, since it appears on the website of an agency of the United States Government, the Centers for Disease Control based in Atlanta (Figure 24.1). The CDC has as its charge the battle against infectious diseases, and it usually is at the forefront when there are outbreaks such as the Zika virus or ebola. However, to advise people about how to deal with a "Zombie Apocalypse" seems to be unusual, to say the least. Will readers actually believe that it is necessary to prepare for such an event? Or will everyone realize that this is merely satire intended to heighten concern generally about how infectious disease can spread? Unfortunately, the CDC, for whatever reason, does not choose

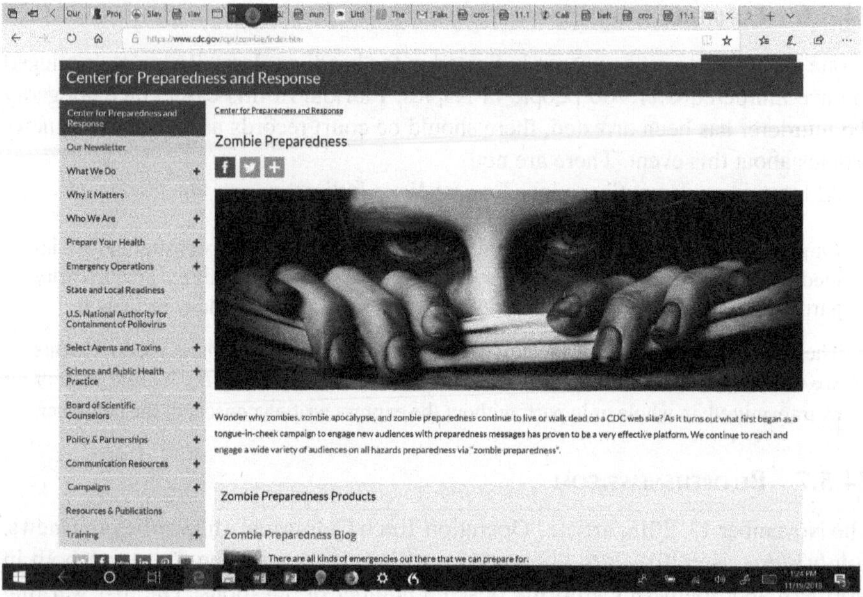

FIGURE 24.1 CDC "Zombie Apocalypse." (https://www.cdc.gov/phpr/zombie/index.htm)

to identify the site as satire, so it would seem this can only be fairly classified as fake news.

24.6 ASSESSING FAKE (OR NOT-FAKE) NEWS

It may be instructive to see what can be learned from human intervention. We will look at a number of these sites to see what "tells" or techniques we can use to identify them as fake.

Fake News Detecting Technique	Explanation
Author Bibliography	See what you can find out about any author indicated. Does that person exist? If so, what are his or her credentials or bibliography?
Authorless	Be suspicious if an article appears and no author is credited.
Comment Section	If the website has a comment section, see if you can determine the credibility of the commenters and the nature of the comments.
Emotional Reaction	How do you feel about the story? Or perhaps, "does it pass the smell test?" Of course, your reaction might depend upon the content.
Expert Testimony	If any experts are quoted, search to see if they really exist; or, if they do, what their qualifications are.
Fact Checkers	There are a number of fact-checking organizations that can be consulted to see if the news item is legitimate. A partial list of these fact checkers will follow.

Continued

Fake News Detecting Technique	Explanation
Grammar	Look for spelling and punctuation errors. It may help to copy the text into Word and run the spellchecker there.
Included Ads	Examine the nature of the ads that might be featured on the suspected site. If the ad indicates you can purchase something online, it may be a scam.
News Outlet	If a news outlet is indicated, if you haven't heard of it, search online for more information.
Other Articles	See if there are other articles on the same topic. If you can't find any, the chances are the story is fake.
Other Sources	Do a search to see if the story at hand is also covered by other media. If it does not appear in the same time frame in a reputable medium, it may very well be bogus.
Publisher "About Us"	In the "About Us" section of the website, see what you can determine about the organization sponsoring the site.
Purpose of the Story	Try to determine the purpose of the story. Is it possible that it is to satisfy an agenda of the publisher, conceivably for political or financial reasons?
Quotes	If the quote is given, search for the source of the quote. See also if you can determine if the person being quoted actually exists or has actual credentials.
Reverse Image Search	If you right-click on an image, you'll find an option to search for the image. Then you should be able to see other websites that have used it and if they are relevant.
Seek Other Experts	If someone in an article is cited as an expert, see how that person is considered by other experts in the same field of inquiry.
Source Check	See if the publisher meets academic citation standards.
Timeliness	Can you verify if the article is recent or perhaps is a copy of something written years before?
URL	When you access the site, examine the URL carefully. On occasion, fake sites have acquired URLs in unlikely countries.
Visual Assessment	Just consider the overall appearance of the site. Once again, you may be able to apply a "smell test."

When seeking fact-checking organizations, you may try these:

FactCheck.org (http://www.factcheck.org)
Politifact (http://www.politifact.org)
The International Fact-Checking Network (https://www.poynter.org/channels/fact-checking)
Snopes.com (http://snopes.com)

REFERENCES

A&E Television Networks. 2017. Marc Antony and Cleopatra. https://www.biography.com/people/groups/mark-antony-and-cleopatra. biography.com. Retrieved July 4, 2017.

Chilton, M. 2016. The War of the Worlds panic was a myth. *The Daily Telegraph*, May 6. https://www.telegraph.co.uk/radio/what-to-listen-to/the-warof-the-worlds-panic-was-a-myth/

Fallon, C. 2014. The shocking, twisted stories behind your favorite nursery rhymes. *Huffington Post*, Nov. 20. https://www.huffingtonpost.com/2014/11/20/nursery-rhymes-real-stories_n_6180428.html

Harrington, H. 2014. Propaganda warfare: Benjamin Franklin fakes a newspaper. *Journal of the American Revolution*, November 10. https://allthingsliberty.com/2014/11/propaganda-warfare-benjamin-franklin-fakes-a-newspaper/

Hoover, J. E. 1946. The enemy's masterpiece of espionage. *Reader's Digest*, 48, April, pp. 1–6.

Kang, C. 2016. Fake news onslaught targets pizzeria as nest of child-trafficking. *The New York Times*, Nov. 21. https://www.nytimes.com/2016/11/21/technology/fact-check-this-pizzeria-is-not-a-child-trafficking-site.html

Kirby, E. J. 2016. The city getting rich from fake news. *BBC News*, Dec. 5. https://www.bbc.com/news/magazine-38168281

MacDonald, E. 2017. The fake news that sealed the fate of Antony and Cleopatra. *The Conversation*, January 13. http://theconversation.com/thefake-news-that-sealed-the-fate-of-antony-and-cleopatra-71287

Soll, J. 2016. The long and brutal history of fake news. *POLITICO Magazine*, December 18. http://www.politico.com/magazine/story/2016/12/fake-news-history-long-violent-214535

Theobald, M. M. 2006. Slave conspiracies in Colonial Virginia. *Colonial Williamsburg Journal*. Winter 2005–2006. http://www.history.org/foundation/journal/winter05-06/conspiracy.cfm. http://www.history.org

White, W. 1992. *The Microdot: History and Application*. Williamstown, NJ: Phillips Publications.

Wikipedia. 2018. List of fake news websites. https://en.wikipedia.org/wiki/List_of_fake_news_websites

PROBLEMS

24.1 Find sources (necessarily an octogenarian+) with a personal recollection of the "War of the Worlds" radio broadcast in 1938. Summarize their recollections.

24.2 Find any recent (post-2016) reference to "Pizzagate."

24.3 Provide an update on fake news arising from Veles, Macedonia. Can you find any other interesting news about that city?

24.4 We have provided eight examples of fake news (or not-fake-news) above. Submit each to the battery of 20 tests indicated in "Fake News Detecting Techniques." What metric would you use to make a final determination of fake news vs not-fake-news? For example, you might say it's fake news if it meets >n of the 20 tests.

25 Potpourri

One may wonder why a chapter in a book on behavioral cybersecurity has a chapter called "Potpourri."

The word potpourri has a number of interesting definitions both in Merriam-Webster and also in Larousse, the standard French-language dictionary, as its origins are in French and it has been transported to English.

The Merriam-Webster definition is as follows:

potpourri

noun

pot·pour·ri | \ ˌpō-pu̇-ˈrē \

Definition of *potpourri*

1. A mixture of flowers, herbs, and spices that is usually kept in a jar and used for scent.
2. A miscellaneous collection: medley a *potpourri* of the best songs and sketches—*Current Biography.*

The French definition, on the other hand (and in translation), is:

Potpourri (literally, "rotten pot"):

- A musical piece created from melodies derived from other works;
- A mixture of several couplets or refrains from various songs;
- A motley mixture of various things; or
- A mixture of fragrant dried flowers.

Plain and simple, the purpose of this chapter is to assemble a number of small nuggets of varying types of information about cybersecurity that hopefully contribute to and expand on the themes found elsewhere in the book. What holds them together is that they illustrate individual points about the various behavioral cybersecurity issues.

Perhaps you can think of them as a mixture of a number of fragrant "flowers, herbs, and tales from cybersecurity."

Some of the issues to follow are:

1. ABCD: A Simple Classification of Cyberattackers
2. The (U.S.) Department of Justice Success in Prosecuting Cyber Criminals: Who's Winning?
3. A Growing Form of Cyberattack: Distributed Denial of Service

4. The Magical Number Seven
5. Good Password Choice
6. Password Creation Videos: Movie Reviews
7. Password Meters

25.1 ABCD: A SIMPLE CLASSIFICATION OF CYBERATTACKERS

When it comes to categorizing the "Bad Guys," we often classify threats as being of type A, B, C, or D to define the level of sophistication of the criminal. We introduced the ABCD principle earlier in Chapter 5, but wish to expand on it here.

Going from the bottom up (from D to A), we can describe each category as in Table 25.1.

Perhaps one way to describe the differences between these four levels of potential threat actors is to describe how these threat levels might be interpreted in a more standard criminal justice or law enforcement environment.

Making the analogy, the D-level street criminal might be epitomized by a person walking along a street where there are a number of parked cars. The potential thief comes to a car where he or she sees a briefcase sitting on the passenger seat. The thief observes the presence of a large rock beside the sidewalk, grabs it, and immediately throws it at the window, shattering it and therefore allowing the person to reach in, take the briefcase, and speed off.

Clearly in this case, there is no planning involved, no background training necessary, and no resources required—other than the rock.

The C-level criminal has some limited skill and knowledge, but does very little planning. Let's go back to the line of cars, and imagine the C-level criminal is also walking down the street. In this case, let's assume the potential thief's objective is actually to steal the car. Since there are many cars on the street, the C thief looks for a target. Obviously, the thief would perhaps look first to see if the car was unlocked, but second, whether locked or not, if he or she noticed a club steering wheel lock, the thief would probably move on to the next car, since the thief's limited skills and resources would be brought to bear in the realization that destroying the club in some way might occupy enough time to expose the thief to capture.

Consider now the same environment for the B-level criminal. It might be observed that there is in fact a well-developed "university" system for the training of the B level criminal: it is called the prison system. Where except in prison can one obtain—without

TABLE 25.1
The ABCD Model for Criminal Behavior Classification

Level of Criminal	Description
D	Smash-and-grab: no skill, knowledge, resources, sees opportunity and acts immediately
C	Some limited skills and resources, but little planning in execution
B	Very knowledgeable, some resources, ability to plan
A	Well-organized team, very sophisticated knowledge, lots of resources, only interested in large targets, can plan extensively

the cost of tuition—the highest level of training in committing criminal acts, with therefore the greatest knowledge but yet a small amount of resources? So now our B-level criminal, back on the streets, knows a great deal about stealing cars but does not have a lot of resources to apply to the task. For example, as with the previous case, the potential criminal might know to stay away from cars with club steering locks, but is also knowledgeable about how to "hotwire" a car to be able to start it up, and might also have done sufficient research to know how often the area might be patrolled by law enforcement so as to be able to estimate how much time he or she might have to actually get away the vehicle. Furthermore, the criminal would probably know about networks where the stolen car might be sold to give the him or her more liquid assets.

The A-level criminal might in fact be the leader or part of an organization that is in the business of receiving the stolen car that might be delivered by the B-level accomplice, but then have the knowledge, resources, organization, and planning to be able to strip down these vehicles and repackage them for future sales.

Translating these automobile examples into the cybersecurity context, we might consider the D-level cybercriminal the person who is pointed to a website that might contain embedded software that can be downloaded and run in order to create some form of attack. In a case such as this, the effort requires no skill or knowledge but is simply acted upon once the malicious software is found and copied. Sometimes the perpetrators of such acts have been called "script kiddies."

The C-level cybercriminal might be one who is able to launch a form of a DDoS attack. For example, by downloading the program "Low Orbit Ion Cannon," one can simply type in the URL and thus launch various types of the DDoS attacks.

The B-level cybercriminal might be someone with a great deal of programming skill, sufficient to create a worm, virus, or ransomware and to launch it to attack some well-chosen target site or sites.

Finally, the A-level cybercriminal might in fact be a government or a vast organization sponsored by government or governments. Certainly examples of A-level attacks might be Stuxnet for its several variants. Stuxnet has been reliably identified as a joint project of both the United States and Israeli governments.

25.2 THE (U.S.) DEPARTMENT OF JUSTICE SUCCESS IN PROSECUTING CYBERCRIMINALS: WHO'S WINNING?

In order to gain an understanding of the relative success of cybercriminals as compared to law enforcement and its ability to deter cybercriminals, for many years the data regarding cyberattacks were very poorly understood. In some ways, this continues to be the case, but there is now a slightly clearer picture, certainly because to some extent there is a more coordinated effort to report on prosecutions for what is considered computer crime.

One major reason for this is that since the beginning of the computer era and the corresponding security concerns with the appearance of viruses, worms, denial of service, and ransomware, the law has simply not involved as rapidly as the technology. For example, consider the following: I may have a very important file or set of files that can realistically be assessed as having a significant monetary value. For the sake of discussion, let us say that this information is worth $1 million. Now, a cybercriminal somehow successfully copies all of this information. Has a theft occurred?

Of course, our instinct would say yes. But if you look at any legal definition of theft, this action cannot be described this way, since the original owner still possesses that information.

For example, the *Peoples' Law Dictionary* defines theft as: "the generic term for all crimes in which a person intentionally and fraudulently takes personal property of another without permission or consent and with the intent to convert it to the takers use (including potential sale)." In many states, if the value of the property taken as low (e.g., less than $500) the crime is "petty theft," but it is "grand theft" for larger amounts, designated misdemeanor or felony, respectively. Theft is synonymous with "larceny." Although robbery (taking by force), burglary (taken by entering unlawfully), and embezzlement (stealing from an employer) are all thought of as theft, they are distinguished by the means and methods used and are separately designated as those types of crimes and criminal charges and statutory punishment.

The term *theft* is widely used to refer to crimes involving the taking of a person's property without their permission. But theft has a very broad political meaning that may encompass more than one category, and multiple degrees, of crimes. Theft is often defined as the unauthorized taking of property from another with the intent to permanently deprive him or her of it, and within this definition, there are two key elements: (a) taking someone else's property and (b) the requisite intent to deprive the victim of the property permanently.

Thus, it is apparent that there is conflict in the definition of the term. On the one hand, obtaining electronic information without the permission of the owner satisfies part of the definition of theft, but on the other hand, the legitimate owner still retains the same information.

Generally speaking, government authorities at most levels have not been able to successfully come to grips with this dichotomy.

So let's look at what we can discover regarding what we know about the prevalence of cybercrime and the success of prosecution.

A number of years ago, the United States Department of Justice began to categorize press releases related to legal actions, usually successful prosecutions, related to a category it called "Cyber Crime."

These releases can be found at the website:

Department of Justice → News,
https://www.justice.gov/news?f%5B0%5D=field_pr_topic%3A3911

At the listing found at the site at the end of September, 2018, a number of releases related to such prosecutions are shown in Table 25.2.

Most of these releases resulted in some successful prosecution. However, reading through the individual cases, the term *virus* is mentioned twice, *ransomware* and *denial of service* mentioned once each, and the term *worm* not at all.

It should also be noted, appropriately, that the United States Department of Justice only reports crimes that fall under the jurisdiction of the U.S. federal government and not similar crimes that might be violations of state or local law. Furthermore, the 2018 data is only for the first 9 months of the year. Nevertheless, aggregating this data would allow us to project an average approximately 30 such cases per year.

TABLE 25.2
Department of Justice Prosecutions for Computer Attacks

Year	Number of Press Releases
2018	29
2017	23
2016	37
2015	21
Total	110

Now, let's look at the other side of the coin, in other words, how well the bad guys are doing. SophosLabs analyzes both malicious attacks but also uses the acronym "PUA" for "potentially unauthorized application," a term that many of us might recognize under the perhaps more appropriate designation of "nuisance."

Nevertheless, Sophos reports that "the total global number of malicious apps has risen steadily in the last four years. In 2013, just over ½ million samples were malicious. By 2015 it had risen to just under 2.5 million. For 2017, the number is up to nearly 3.5 million." The vast majority are truly malicious with 77% of the submitted samples turning out to be malware.

This is a worldwide sample; however, at least with respect to ransomware, the U.S. data represent 17.2% of the global figure. Thus, we could estimate reasonably that for 2017, the prevalence of grants and more attacks in the United States would exceed 500,000.

And, of course, ransomware attacks form only a portion of the complete number of malicious attacks, but nevertheless, just compare: half a million attacks per year versus 30 prosecutions.

It would lead one to believe that the odds are pretty good for the cybercriminal.

25.3 A GROWING FORM OF CYBERATTACK: DISTRIBUTED DENIAL OF SERVICE

Distributed denial of service attacks are coordinated efforts by human or machine to overwhelm websites and, at a minimum, to cause them to shut down. The use of this type of malicious software has grown exponentially in the past decade, and despite considerable research, it has proven very difficult to identify, detect, or prevent such attacks. On the other hand, increases in traffic at websites may not be the result of a DDoS attack, but a legitimate increase in demand for the Web service.

The cybersecurity community has focused more attention on the types of malware that require at least a measure of programming skill to create and distribute: viruses, worms, Trojan horses, bots, and so on. But now, it has been well documented that there is exponential expansion in the number and intensity of DDoS attacks, and it has become a major concern for system administrators and their organizations.

In the 2011 publication of the World Infrastructure Security Report (WISC, 2011), it was noted that the reported increase in DDoS attacks had been multiplied by a factor of 10 since the first year of the study in 2004, and that ideologically motivated "hacktivism" and vandalism have become the most readily identified DDoS attack motivations. In the 2012 publication of the same report (WISC, 2012), the trend toward political motivation was quantified as follows: "35% [of internet-connected organizations] reported political or ideological attack motivation … 31% reported nihilism or vandalism as attack motivation." In the past 7 years, the number and intensity of DDoS attacks have been increasing at an exponential rate. Again, as reported by the WISC (2011), in 2011, there was a significant increase in flood-based attacks in the 10-Gbps (Gigabits per second) range.

DDoS attacks have entered the mainstream, and they have been described as the "new normal." In its simplest form, a DDoS attack is a coordinated set of requests for services, such as a Web page access. These requests may come from many nodes on the Internet, either by human or electronic action, and the requests require resource utilization by the site under attack. Furthermore, it is now possible with certain types of attacks to participate in them by doing nothing more than entering a URL and clicking, thus increasing the number of users capable of participating in a DDoS attack to include virtually everyone who has internet access. For example, the Low Orbit Ion Cannon (LOIC) is easily accessible on the Internet, and use of this software to initiate or participate in a DDoS attack only requires typing in the website name. Often the DDoS attack only has as its objective stimulating so much traffic on a website that the bandwidth capacity of the host system is overwhelmed and it crashes. The recovery time for an organization whose website has crashed may range from a few hours to a few days (Figure 25.1).

Despite considerable research on DDoS attacks over the past decade, there is little coordinated information about specific attacks. This is understandable because attacked sites may not wish to publicize the fact that such attacks have occurred. More

FIGURE 25.1 Low Orbit Ion Cannon.

and more often, however, an attacking group may publicize that it has brought down a site (Dominguez, 2011; Mutton, 2011; Poulson, 2010; Singel, 2011). However, a DDoS attack might be indistinguishable from a sudden influx of requests because of a specific event. For example, news sites might have an extraordinary increase in legitimate requests when a significant event occurs—the death of a celebrity, for example, or a ticket brokerage service may be flooded when popular tickets go on sale. We have used the term "anomalous event" to describe the condition where there is a rapid increase in demand for Web service, and this could occur either under a DDoS attack or legitimate heavy traffic. Thus, the ability to monitor more precisely such anomalous events could be a useful tool in developing better prevention or detection mechanisms.

This research is based on an attempt to characterize DDoS attacks using data supplied on the internet by the internet traffic company, alexa.com, and also other sources (Alexa, 2019). Because of the wealth of data available through this service, we have attempted to build a database that will yield information about the behavior not only of DDoS attacks, but other anomalous behavior in traffic at websites. Figure 25.2 shows the result of the alexa.com report on "reach," the estimated percentage of global internet users who visit doj.gov, the public website of the U.S. Department of Justice. The parameters in this traffic report are time (t), on the horizontal, which we have recorded in days, and reach (r), which, according to Alexa.com, represents a percentage of the overall internet traffic.

In particular, Figure 25.2 shows the level of traffic on the Department of Justice website over a 6-month period in 2011/2012.

It is easy to see that there is a considerable anomaly in this traffic in the latter part of January 2012. We described earlier in this book (Chapter 1) that there had been, immediately after WikiLeaks released many documents originally received from U.S. Army Private Chelsea Manning, a "cyberwar" that began with various attackers attempting to bring down WikiLeaks with a DDoS attack. When this failed,

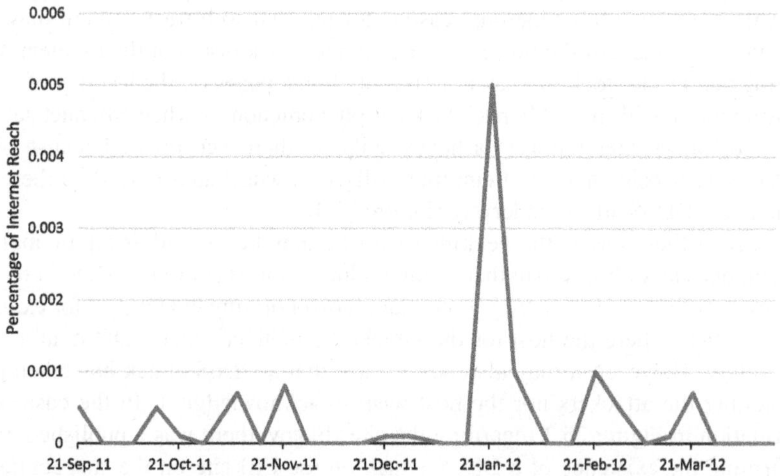

FIGURE 25.2 Internet traffic on the Department of Justice website (Oct. 2011–Apr. 2012).

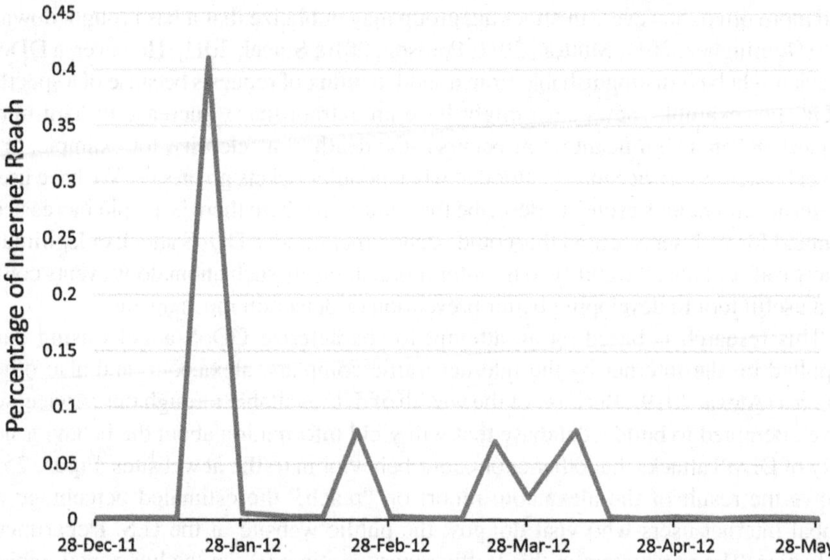

FIGURE 25.3 Alexa.com traffic report for fbi.gov website (Oct. 2011–Apr. 2012).

supporters of WikiLeaks launched in retribution a DDoS attack and brought down the Department of Justice website, as indicated in Figure 25.2.

That DDoS attack was also replicated in the same period against the FBI website, as is demonstrated in Figure 25.3.

We have examined the traffic reports for many websites and have used a protocol that a particular traffic report yields an anomalous event if (1) the traffic increase or spike at a given point in time is isolated in an appropriate interval, and (2) the height of the traffic at the spike is greater than the height of the traffic in any appropriate interval. There may be numerous reasons for unusual web traffic: there may be a cycle in the business of the host site, for example, stock prices at the moment of the opening bell in the stock market; at university home pages on the last day of course registration; or with the "Michael Jackson phenomenon"—when Michael Jackson died, most news sites reported a heavy spike in their web traffic because of the widespread curiosity in users attempting to discover what had occurred; or there may be an actual DDoS attack underway (Figure 25.4).

In any of these cases, the peaking of traffic may be viewed as an anomaly; In general, one can only confirm that an anomalous event is actually a DDoS attack or a "heavy traffic" event if there is secondary, corroborating evidence—for example, a news article where the host (or the attacker) announces that a DDoS attack has taken place. However, it may also be the case that a DDoS attack has taken place and neither the attackers nor the host wish to acknowledge it. In the case of the information in Figure 25.3 regarding the site fbi.gov, there was a published article confirming the existence of a DDoS attack on the FBI site on the suspect date in January 2010 (Mick, 2011).

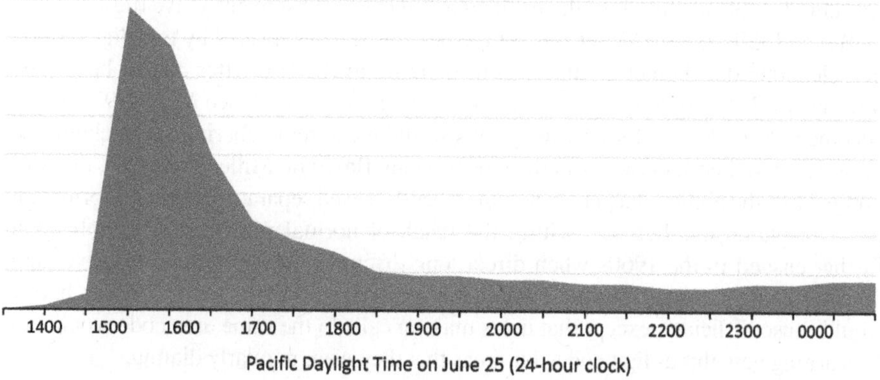

FIGURE 25.4 Internet traffic report on June 25, 2009 (Michael Jackson's death).

At the time of this writing in 2018, the open availability of the Alexa data has been considerably restricted. However, what is available now still presents an interesting picture of internet traffic on certain sites.

The original reporting on this data came as a result of a joint research team consisting of students and faculty from Howard University in Washington, DC, and the Universidad Santo Tomás in Santiago, Chile, working together to develop this research. The results were described in Banks et al. (2012).

25.4 THE MAGICAL NUMBER SEVEN

Long before computers became widespread, a research paper written by George Miller, a psychologist at Harvard University, demonstrated a number of principles that turned out to have an enormous impact on cybersecurity many years later.

The paper, called "The Magical Number Seven, Plus or Minus Two" (Miller, 1956), had an influence originally on the telephone system design for assigning numbers and subsequently on the design for the rules for formulating computer passwords.

Miller was able to demonstrate that a normal human's capacity for remembering random pieces or "chunks" of information centered around seven distinct pieces of information.

Going back to the 1940s, most telephone systems had phone numbers consisting of four digits, then five digits and subsequently a two-letter code indicating the neighborhood of the location of the telephone.

There are remnants of telephone numbers assigned under that system from older film, television, and music. The oft-repeated television sitcom "I Love Lucy" used a fictional telephone number for the Ricardo residence, MU 5-9975 (MU for Murray Hill was an actual telephone exchange region in Manhattan); a popular song of the 1940s was "PE 6-5000" (read PEnnsylvania 6-5000 or "Pennsylvania-six-five-thousand"); and a popular movie of the 1950s starring Elizabeth Taylor was "Butterfield 8" for a fictional telephone exchange BU 8-xxxx.

The advantage of that system was that essentially the user only had to remember six "chunks" of information: the exchange name and subsequently five digits.

By the 1950s, the two-letter code at the beginning was replaced by two digits (often the telephone dial numeric values corresponding to the two-letter code—hence the Murray Hill MU would become "68," so that the Ricardo residence MU 5-9975 would become 685-9975). But for new users, this would mean remembering seven chunks of information rather than the six in the older system. Based on Miller's research, this still implied that the ability of a person to remember the seven separate pieces of information for a telephone number was within the reach of normal memory. A complication further ensued in the 1960s when direct long-distance dialing became possible, and now telephone numbers became 10-digit numbers. According to Miller's research, this would cause difficulty, except that users making calls to the same area code would not be learning new three-digit codes for areas that they were regularly dialing.

The essence of Miller's research and subsequent observations was a number of prior experiments asking subjects to discriminate between or identify different objects as the number of objects was gradually increasing. In virtually all of the research he reported, the ability of the subjects to identify the precise number of objects observed tended to decline rapidly when the number presented was near seven. A few of the examples in his research were:

1. Pollack (1953) asked listeners to identify auditory tones by assigning numerals to them. The listener responded with a numeral when each tone was played. When the experiment was established with only 2 to 4 different tones, very few incorrect identifications occurred. As the number of tones increased to between 5 and 14, the listeners made many mistakes. The correct number of identifications reached a level of approximately six; the author called this the channel capacity of the subject for absolute judgments of pitch.
2. Garner (1953): In this example, the test was on loudness of sounds. He spaced a number of tones in intensity between 15 and 110 dB (decibels), and used between 4 and 20 different intensities with each subject. Again the range of the subject's ability to recognize the difference in loudness was about five perfectly describable alternatives.
3. Beebe-Center, Rogers, and O'Connell (Beebe-Center et al., 1955) dealt with taste intensities. In this case, the responses had to do with absolute judgments on the concentration of salt solutions. These concentrations were between 0.33 to 34.7 g of salt per 100 mL (milliliters) of tapwater. In this case, the ability to recognize peaked at around four distinct concentrations.
4. Hake and Garner (1951): The experiment was to ask observers to identify the position on a number line as an estimate of a marker at an observed position on the number line. In this case, there were markers at only 5, 10, 20, or 50 different positions. With this experiment, the average number of correctly estimated positions tended to average around nine.

In the context of behavioral cybersecurity, the importance of Miller's findings relates to the construction of password systems. Since one of the greatest weaknesses in password systems is the inability of the user to successfully recall a password, it is critical in the design of such systems that the user be able to recall the password that he or she has established. If the user is unable to recall a password, then one of two outcomes are the most likely. First, the user will go through a process to change a password, which will be at the least an annoyance and at the most a significant irritation and frustration, or second, in order to avoid the frustration, the user might write down the password, exposing it to visibility by a third party who might have malicious intent.

Thus, we can conclude from Miller's research that the likelihood of being able to remember a password is best if it is on the order of seven characters. It should also be noted that simply using a seven-letter word that might be found in a dictionary also exposes the account to what is called a "dictionary attack," whereby the intruder, who can obtain a copy of the dictionary, can test each word until finding a "hit."

25.5 GOOD PASSWORD CHOICE

How to create good passwords is a long-standing, but nevertheless still perplexing, problem for computer users everywhere. Undoubtedly this problem will remain until such time as the community of computer users determines that passwords are not an appropriate test for authentication—for example, to be replaced by biometrics (what you are) or physical devices (what you have). But of course, this might never happen.

In any case, we've tried in this book to outline some of the concerns about password choice. One solution—as we will discuss in the next section—is the use of "password meters," in other words, software that will indicate the strength of a potential password. However, experience with password meters has shown that in general and at present, they tend to be quite unreliable.

As also has been discussed in an earlier section ("The Magical Number Seven"), when, as in most cases, the user must rely upon a password that must be committed to (human) memory, one might say, well let's avoid that problem by (a) storing that password somewhere on our computer or (b) writing that password down.

Of course, both of these approaches expose the password to an attacker who either gains access to our computer or can rifle through our desk drawer.

On the other hand, if one assumes that we must commit a password to memory, then we must be confident that our memory is sufficient to contain such a password—or, in general in these modern times, contain the multiple passwords that we need for not only our computing account, but access to many other websites such as our bank account, our accounts with bills that we may have to pay electronically, or online vendors where we may purchase various merchandise.

With the use of multiple passwords, of course our human memory requirements multiply as well.

Various solutions have been proposed:

- Random strings of characters, for example, vursqpl. Bad idea! Who could ever remember such a random string?
- Very long strings, for example, for Abraham Lincoln: fourscoreandseven yearsagoourforefathers. Certainly memorable, but for one, Abe's fingers might get tired of typing that on every occasion, and also, it might be more easily guessed by anyone who studied Lincoln.

- *Passphrases*: rather than the long string as indicated above, create a short expression, for example, from the first letters of each word in the phrase. Thus, Abe's long password would become fsasyaof (or, in one variation, fsa7yaof). Now the password is short enough to remember what seems like a random string, but the user has in his or her human memory the more famous long expression and remembers the rule of just taking the first character of each word in the phrase. The defect here comes if the phrase chosen by the user is easily one that one could conclude from greater knowledge of the user.

It is also possible to become a "movie reviewer," to learn about other views of the proper construction of passwords. There are several that be found on YouTube, for example.

Identity Theft Manifesto:
https://www.youtube.com/watch?v=LOJnBI1HDn4

Kelsey Dobish and Trey Shafto:
https://www.youtube.com/watch?v=jWablPsy6ng

Endsight Corporation:
https://www.youtube.com/watch?v=UnmdxReVoNc

Graham Cluley of the Sophos Corporation:
https://www.youtube.com/watch?v=VYzguTdOmmU

Our students provided "reviews" of a number of these sites, as we will describe in the next section.

The suggestion that we tend to use is to choose a relatively short but memorable event, so that it is indelibly burned into our consciousness, but also one that is not well known to persons who wish to research our histories. Let's say that many years ago, I had a memorable encounter with a person, place, or thing—let's say it was "Penobscot," which is actually a river in the state of Maine.

Furthermore, in this case, I had told no one of having spent some time along that particular river—so no one who knew me would make such an association. So now this is a password I am unlikely to ever forget—but, unfortunately, there are dictionaries of place names for river names that could be used in a random attack to obtain my password. In order to still use "Penobscot" but isolate it from such attacks, I might insert some other character into the password so that simply by testing for the password with a dictionary of river names, we might deflect such an attack by using as the password, for example, penob7scot.

In addition to this one example, it's prudent for any user to develop a list of such words that are highly memorable and yet, when modified in some fashion such as the above, will frustrate any attack just based on the use of some well-known set of dictionary words.

25.6 PASSWORD CREATION VIDEOS: MOVIE REVIEWS

It is clear that the major issue in the use of passwords for authentication relates to the challenges to human memory in deeply embedding the choice of password or choice of several among many passwords required for a particular user. We have addressed above some principles for good password choice.

In addition, through the ubiquitous YouTube, many individuals and organizations have filmed and posted their own prescriptions for how to create good passwords. At last count, one can find at least 14 separate YouTube videos giving explanations as to methods for making good password choices.

We posed the problem to our students to assess the quality of information and the quality of the videos that are readily available through the YouTube site.

We provide a summary of the views of our students in the critique of a number of these videos (Figures 25.5 through 25.8).

FIGURE 25.5 Choosing passwords video (Identity Theft Manifesto) (https://www.youtube.com/watch?v=LOJnBI1HDn4).

(From Identity Theft Manifesto)

- This video gave a lot of good information. It talks about how guessable information is not a good password and also talks about how the combination of uppercase and lowercase as well as symbols, letters, and numbers are good passwords. PH
- However, they tell you to write it down, which is not such a good idea.
- Thumbs down. Video was way too long. The video seemed poorly made (from song to video editing and everything else in between). OO
- She exaggerates too much.
    - Her tone was annoying.
    - I feel as if the information too obvious.
    - The background music is distracting. SS

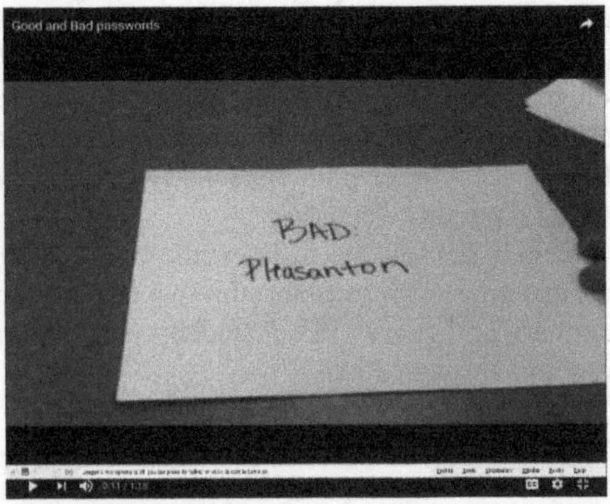

FIGURE 25.6 Choosing passwords video (Kelsey Dobish and Trey Shafto) (https://www.youtube.com/watch?v=jWablPsy6ng).

(From Kelsey Dobish and Trey Shafto)

- This video suggests the best idea that I've heard so far in terms of passwords: to use phrases that are relevant/memorable to you. PH
- Video was not the best but the content was pretty good. It did supply its viewers with a decent way to come up relatively safe passwords. OO
- Video is shaky. SS

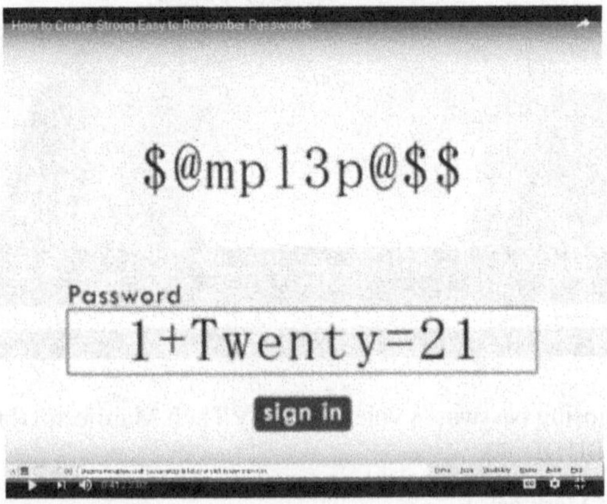

FIGURE 25.7 Choosing passwords video (Endsight Corporation) (https://www.youtube.com/watch?v=UnmdxReVoNc).

(From the Endsight Corporation)

- The creator provided a decent way of creating relatively safe passwords. What I thought was funny: using the website provided to find the strength of a password, the phrase "I was born in San Francisco in 1972." would take a computer "402 QUINDECILLION YEARS" to crack while "IwbiSFi1972." would take "63 THOUSAND YEARS" to crack. OO
- A GREAT VIDEO!! SS

FIGURE 25.8 Choosing passwords video (Sophos Corporation) (https://www.youtube.com/watch?v=VYzguTdOmmU).

(Presented by Graham Cluley of the Sophos Corporation)

- Video was well made. Like the second and third videos, it provided a similar way of creating passwords. OO
- Useful information.
- His accent might be confusing to some. SS

(Reviews courtesy of students Portia Herndon, Osi Otugo, and Shanay Saddler.)

25.7 PASSWORD METERS

Another approach to the creation and use of secure passwords is through one of a number of so-called "password meters" that are available at various sites on the Internet.

In particular, we have examined five of these candidates for password meters. In each case, the meter is available through a particular website, the user is encouraged to enter a test password, and a report is generated for the user as to the password meter's judgment of the strength of the password. They are shown in Table 25.3.

Unfortunately, what we have too often discovered is that the strength of password judged by any one of the test password meters may vary completely from the strength

TABLE 25.3
Websites for Password Meters

Password Meter Designation	Name
A	https://passwordmeter.com
B	https://lastpass.com/howsecure.php
C	https://my1login.com/resources/password-strength-test/
D	https://thycotic.com/resources/password-strength-checker/
E	https://howsecureismypassword.net/

of the same password as judged by one or several of the other candidates for password meters.

One way of testing the validity for consistency of a proposed password meter is to submit a number of prospective passwords to each meter candidate and examine the consistency of the results.

For a test, suppose we use the password choices in Table 25.4.

Merriam-Webster Definition of *onomatopoeia*

1. The naming of a thing or action by a vocal imitation of the sound associated with it (such as *buzz, hiss*) also: a word formed by onomatopoeia In comic books, when you see someone with a gun, you know it's only going off when you read the *onomatopoeias.*—Christian Marclay
2. The use of words whose sound suggests the sense: a study of the poet's *onomatopoeia*

The results of these proposed passwords on each of the five password meter sites (A–E) as indicated above are as found in Table 25.5. We will return to this in "Hack Lab 4," Chapter 26 (Table 25.5).

TABLE 25.4
Test Passwords to Evaluate Password Meters

Test Password	Feature
1111111111111111111	A rather simple but very long string. A nuisance to type in.
Penob7scot	A word found in some types of dictionaries, for example, a dictionary of US place names, with the insertion of a number to defeat a dictionary attack.
x3p9q!m	A seemingly random string, but probably difficult to remember.
brittttany	A person's name, with a certain letter duplicated (T). Probably easy to remember.
onomatopoeia	A long word, but in most dictionaries.
aBc123xYz	Follows a pattern, but the mixture of letters, numbers, and capitals could prove strong and in this case probably easy to remember.

TABLE 25.5
Comparison of Password Meters

Test Password	*A*	*B*	*C*	*D*	*E*
1111111111111111111	6 Very Weak 0%	2 Weak	6 Very Weak 0.04 seconds	3 2 weeks	1 79 years
Penob7scot	2 Strong 63%	2 Moderately Strong	2 Strong 7 months	1 3 years	2 8 months
x3p9q!m	3 Good 54%	4 Weak	1 Strong 8 months	6 1 minute	6 22 seconds
britttany	5 Very Weak 8%	2 Moderately Strong	5 Weak 16.05 minutes	5 4 hours	5 59 minutes
onomatopoeia	4 Very Weak 13%	1 Very Strong	3 Medium 13 hours	2 4 months	3 4 weeks
aBc123xYz	1 Strong 76%	4 Weak	4 Weak 40.01 minutes	3 2 weeks	4 4 days

Note: Each password meter normally indicates both a non-numeric measure of strength of a password, for example "weak," "moderately strong," "good," etc. Also the meter indicates a numeric value representing the projected length of time to find the password."

REFERENCES

Alexa. 2019. http://www.alexa.com

Banks, K. et al. 2012. DDoS and other anomalous web traffic behavior in selected countries. *Proceedings of IEEE SoutheastCon 2012*, March 15–18, 2012, Orlando, Florida.

Beebe-Center, J. G., Rogers, M. S., and O'Connell, D. N. 1955. Transmission of information about sucrose and saline solutions through the sense of taste. *The Journal of Psychology*, 39, 157–160.

Dominguez, K. 2011. Bitcoin mining botnet found with DDoS Capabilities. *Malware Blog*, September 4. http://blog.trendmicro.com/bitcoin-mining-botnet-found-with-ddoscapabilities/

Garner, W. R. 1953. An informational analysis of absolute judgements of loudness. *The Journal of Experimental Psychology*, 46, 373–380.

Hake, H. W. and Garner, W. R. 1951. The effect of presenting various numbers of discrete steps on scale reading accuracy. *The Journal of Experimental Psychology*, 42, 358–366.

Mick, J. 2011. LulzSec downs CIA's public site, appears to be subject of framing attempt. *Daily Tech*, June 15, http://www.dailytech.com/LulzSec+Downs+CIAs+Public+Site+Appears+to+be+Subject+of+Framing+Attempt/article21916.htm

Miller, G. 1956. The magical number seven, plus or minus two: Some limits on our capacity for processing information. *Psychological Review*, 63, 81–97.
Mutton, P. 2011. LiveJournal under DDoS attack. *Performance*, April 4, http://news.netcraft.com/archives/2011/04/04/livejournal-under-ddosattack.html
Pollack, I. 1953. The assimilation of sequentially encoded information. *American Journal of Psychology*, 66, 421–435.
Poulson, K. 2010. Cyberattack against WikiLeaks was weak. wired.com, November.
Singel, R. 2011. FBI goes after anonymous for pro-WikiLeaks DDoS attacks. Wired.com, January 28. https://arstechnica.com/tech-policy/2011/01/fbi-goes-after-anonymous-for-pro-wikileaks-ddos-attacks/?comments=1
WISC. 2011. *World Infrastructure Security Report 2010*. Chelmsford, MA: Arbor Networks.
WISC. 2012. *World Infrastructure Security Report 2011*. Chelmsford, MA: Arbor Networks.

PROBLEMS

25.1 Use the ABCD method to classify the following:

Crime/malware	A, B, C, or D
Pickpocket	
Fabrication of student grades	
The Morris worm	
Script kiddies	

25.2 In the Department of Justice news website (https://www.justice.gov/news?f%5B0%5D=field_pr_topic%3A3911), find the number of cases in 2017–2018 involving:

a. Hacking
b. Phishing
c. DDoS
d. Ransomware

Is each on the increase, decline, or nonexistent (according to DoJ)?

25.3 Find a recent example where internet traffic indicates a spike in traffic to a certain website. Can you determine if it is DDoS or the "Michael Jackson effect"?

25.4 Do your own reviews of the password choice videos. Add any other relevant videos you might find beyond the four discussed above.

26 Hack Lab 4 *Contradictions in Password Meters*

Throughout this book you will discover a number—in particular, four—of what we call "Hack Labs." These labs are designed to give students practical experience in dealing with a number of cybersecurity issues that are of critical concern in the protection of computing environments.

The other purpose for these labs is that it is not necessary, but there could be a supportive physical computer lab to carry out these projects. They can also be done on a student's own computing equipment and do not have to be done within a fixed lab period.

When these have been offered by the authors, they have usually allowed the students a week to carry out the research and submit the results.

26.1 HACK LAB 4: PASSWORD METERS

This Hack Lab is meant to have students judge the consistency and effectiveness of a number of password meters that can be found on the internet.

Password Meter Designation	Name
A	https://passwordmeter.com
B	https://lastpass.com/howsecure.php
C	https://my1login.com/resources/password-strength-test/
D	https://thycotic.com/resources/password-strength-checker/
E	https://howsecureismypassword.net/

The following is one sample result using sample passwords generated by the authors.

Test Password	*A*	*B*	*C*	*D*	*E*
1111111 1111111 111111	6 Very Weak 0%	2 Weak	6 Very Weak 0.04 seconds	3 2 weeks	1 79 years
Penob7scot	2 Strong 63%	2 Moderately Strong	2 Strong 7 months	1 3 years	2 8 months
x3p9q!m	3 Good 54%	4 Weak	1 Strong 8 months	6 1 minute	6 22 seconds
brittttany	5 Very Weak 8%	2 Moderately Strong	5 Weak 16.05 minutes	5 4 hours	5 59 minutes
Onomatopoeia	4 Very Weak 13%	1 Very Strong	3 Medium 13 hours	2 4 months	3 4 weeks
aBc123xYz	1 Strong 76%	4 Weak	4 Weak 40.01 minutes	3 2 weeks	4 4 days

The formula that can give you a single datum comparing all solutions would be as according to the computation of DISTANCEi below.

Suppose the indices we use are the indices in the Test Password array. For the example above.

Take the square of the differences of all elements in each row, and call the total DISTANCE. The columns are called A, B, C, D, E and the rows 1, 2, 3, 4, 5, 6:

$$\begin{aligned}\text{DISTANCEi} = {} & (\text{Bi}-\text{Ai})^2+(\text{Ci}-\text{Ai})^2+(\text{Di}-\text{Ai})^2\\ & +(\text{Ei}-\text{Ai})^2+(\text{Ci}-\text{Bi})^2+(\text{Di}-\text{Bi})^2\\ & +(\text{Ei}-\text{Bi})^2+(\text{Di}-\text{Ci})^2+(\text{Ei}-\text{Ci})^2\\ & +(\text{Ei}-\text{Di})^2\end{aligned}$$

Repeat this computation for i = 1, 2, 3, 4, 5, 6.

Showing only the calculation for DISTANCE1:

$$\begin{aligned}&(2-6)^2+(6-6)^2+(3-6)^2+(1-6)^2+(6-2)^2+(3-2)^2\\ &+(1-2)^2+(3-6)^2+(1-6)^2+(1-3)^2\\ &=16+0+9+25+16+1+1+9+25+4=106.\end{aligned}$$

PROBLEM

26.1 Use your own password choices to test this algorithm. See if you can find password choices that maximize the distance function described above.

27 Conclusion

You have now arrived at the end of our story—for now. We hope you will find it an interesting challenge to carry on your learning about these increasingly important issues arising from cybersecurity and the behavioral sciences. We believe this compound subject involving both the knowledge of computing disciplines as well as behavioral science disciplines is still in its infancy. In particular, there are paths that can lead to important future research understanding based on the following.

27.1 PROFILING

We have provided a few examples of profiling techniques, but the reader can find many other examples of accounts of other attacks that have taken place in recent years. The Sony Pictures example we have used is interesting because it touches on a number of varieties of subjects and motivations. But the reader might find it equally of interest to analyze the cases of WannaCry, Petya, fake news incidents, and numerous others.

27.2 SOCIAL ENGINEERING

Social engineering provides a continuing and important avenue for the development of hacking approaches to unlawful entry. A student studying the success of social engineering methods can approach it from two perspectives: first, understanding the exposure that a careless user provides to a hacker or "dumpster diver" in trying to guess a certain password. On the other hand, one can also be cognizant of the security provided when password methods are developed with great care. This book has also provided an introduction to the utility or futility of what we now call password meters, and the field is open for persons who will potentially design criteria for such meters in a more rigorous fashion.

Persons interested in this area of study should also take into consideration the challenges of trying to move from what we have referred to as one-factor authentication, such as passwords or biometrics, to multifactor authentication, both from the perspective of the security provided but also the difficulty in migrating from the world's supply of password system users to a multifactor environment.

27.3 SWEENEY PRIVACY

The results of Sweeney's research (Chapter 6) showing that almost 90% of users in the United States can be determined uniquely by three usually widely available pieces of information are shocking to many people. Further research in this area might consider the level of security provided by other forms of identification. Future research by the authors of this book will illustrate that the data described in Sweeney's research make the issue of privacy even less secure in a number of countries, while similar data do not pose as much of a threat in certain other countries.

27.4 UNDERSTANDING HACKERS

We have tried to provide a number of examples of the psychology of certain hackers that have been willing to describe themselves in the literature. But this is a field that is wide open to further study. There has not been a great deal of literature on the analysis of motivations of greater number of identified hackers, and even less on the understanding of behavioral tendencies of hacker organizations.

27.5 GAME THEORY APPLICATION TO PROFILING

Game theory has a considerable body of scholarship in general but has had very little application to the field of cybersecurity. We believe that in all probability this is because researchers in cybersecurity problems have not tended to take a quantitative approach to the research. This is an approach we have tried to emphasize throughout this book, and although we realize that in a number of cases attempting to quantify issues related to behavior can be challenging, nevertheless we feel that this is not the reason to avoid a quantitative approach altogether, such as we might use in game theory techniques.

27.6 TURING TESTS

The Turing test remains a fascinating example, bequeathed to us by Alan Turing, of a methodology for distinguishing not only between human and computer, but also, as we have projected, between different groups of humans (or machines) in our environment. The specific example we have pursued is related to the determination of gender, as we have called it, the "gender Turing test." But readers may also be interested in developing an "age Turing test," a "professional training Turing test," or conceivably others.

27.7 CRYPTO AND STEGO

An area that could bear very significant research is the concept of developing security measures that use a hybrid approach involving both cryptography and steganography. It would seem that the importance of developing such an approach would be to use the aspect of steganography that tests human behavior as a tool supplemental to that provided by cryptography.

27.8 BEHAVIORAL ECONOMICS

This field has become a major area of research in economics, although many would also argue that it is as significant a subdiscipline in psychology as it is in economics. In our case, there has been very little study about how this may be applied to issues of cybersecurity, so this field is in our minds wide open.

27.9 FAKE NEWS

This is in many ways a new area on the scene. Although, as we described in Chapter 23, we can track "fake news" virtually to the dawn of human history, nevertheless the distribution of such news has of course been empowered by our current technological area, in particular with the use of the internet. The question that we have suggested for further research is the development of human-developed or artificial intelligence techniques for the detection of what we call fake news, and this is not only a wide-open area but also of substantial interest to major organizations in the computing industry.

27.10 PASSWORD METERS

We have mentioned the, generally speaking, lack of success in developing meters to measure password strength. There is a question as to whether this field can be improved or whether there are standard measures that could apply to all situations to determine password strength or the lack of it.

27.11 NEXT STEPS

The authors are very interested in encouraging the development of coursework and research in this emerging field of behavioral cybersecurity. As a consequence, we will be very pleased to hear from many students and professors involved with either coursework or research in this new and promising area.

We encourage contact of any form and suggest that the best means of informing us of your progress is the email address: waynep97@gmail.com

Index

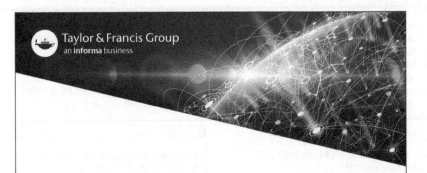